Reimagining Alternative Technology for Design in the 21st Century

Reimagining Alternative Technology for Design in the 21st Century presents a new approach to design that harnesses still-valuable alternative, traditional and abandoned technologies alongside the creation of new ones to address contemporary global problems. It focuses on design opportunities that reduce energy and material consumption to tackle issues such as climate change and pollution in industrialized economies.

The book takes the reader on a journey surveying different facets of human activity to identify underused and discarded technologies that could be indispensable today. It critically addresses newer approaches to design and technology by comparing them to existing alternatives, unpacking examples including air conditioning with smart thermostats, electric lighting, durable reusable products, domestic maintenance tools and methods of transportation.

Written for practicing designers and students in industrial design, architecture, sustainable design and human-centered design, this book provides new ideas and tools for creating more useful, energy-and-resource-efficient product designs and systems.

Brook S. Kennedy is an industrial designer and Associate Professor in the School of Architecture, Arts and Design at Virginia Polytechnic Institute and State University (Virginia Tech), US. There his research focuses on topics in sustainable design, materials and biodesign. Prior to Virginia Tech he spent 15 years in full time practice, including as an Associate Director at Inclusive Design pioneer Smart Design in New York City, where he helped Fortune 500 clients and start-up companies embrace the value of human-centered design processes. He has been honored with numerous international design awards from the Industrial Design Society of America, International Forum (iF), Red Dot (Best of the Best), Spark!, European Product Design Awards and the Chicago Athenaeum among others. His work has been covered in the *New York Times*, *PBS Newshour*, *Fast Company*, *NPR* and the *Washington Post* and has been exhibited internationally. He holds more than 20 utility and design patents and has published and exhibited internationally.

"The age of buildings as 'machines for living in' is ending. We live in an increasingly unpredictable world. 'Fit for purpose' building in the 21st century must cope with evermore extreme weather events, grid failures, finite natural resources and soaring costs of living. Buildings, and the products and furnishings inside them, must be durable, repairable and adaptable. Alternative technologies re-connect people with the free, natural energy flows of the ecosystems they occupy, enabling them to do more for less, educating them to protect against and exploit those powerful flows where useful. This book is a great read for all who are keen to survive and thrive in a different future."

Susan Roaf, *Professor of Architectural Engineering at Heriot-Watt University, UK*

"Sometimes the best idea is one that's tried, true and ready for a comeback. Kennedy makes a brilliant argument for innovative revivals of ideas, products and goods whose time has once again come. A must-read for anyone looking for a more sustainable future!"

David Sax, *journalist based in Canada, and author of the best-selling book,* Revenge of the Analog

Reimagining Alternative Technology for Design in the 21st Century

BROOK S. KENNEDY

Routledge
Taylor & Francis Group

LONDON AND NEW YORK

Designed cover image: SHORPY

First published 2023
by Routledge
4 Park Square, Milton Park, Abingdon, Oxon OX14 4RN

and by Routledge
605 Third Avenue, New York, NY 10158

Routledge is an imprint of the Taylor & Francis Group, an informa business

© 2023 Brook S. Kennedy

The right of Brook S. Kennedy to be identified as author of this work has been asserted in accordance with sections 77 and 78 of the Copyright, Designs and Patents Act 1988.

British Library Cataloguing-in-Publication Data
A catalogue record for this book is available from the British Library

Library of Congress Cataloging-in-Publication Data
Names: Kennedy, Brook S., author.
Title: Reimagining alternative technology for design in the 21st century /
    Brook S. Kennedy.
Description: Abingdon, Oxon ; New York, NY : Routledge, [2023] | Includes
    bibliographical references and index.
Identifiers: LCCN 2022053885 (print) | LCCN 2022053886 (ebook) |
    ISBN 9780367410230 (hardback) | ISBN 9780367410223 (paperback) |
    ISBN 9780367814304 (ebook)
Subjects: LCSH: Technological innovations. | Appropriate technology. |
    Technological forecasting. | Sustainable design. | Product design. |
    System design.
Classification: LCC T173.8 .K458 2023 (print) | LCC T173.8 (ebook) |
    DDC 601/.120286—dc23/eng/20230208
LC record available at https://lccn.loc.gov/2022053885
LC ebook record available at https://lccn.loc.gov/2022053886

ISBN: 978-0-367-41023-0 (hbk)
ISBN: 978-0-367-41022-3 (pbk)
ISBN: 078 0 367 81130 4 (obk)

DOI: 10.4324/9780367814304

Typeset in Univers LT Std
by Apex CoVantage, LLC

Contents

List of Figures

Acknowledgements

I would like to thank the many people who have contributed to this project. Above all, I would like to thank my wife Anna, who has always supported me and inspired countless discussions about the book's topic. I would also like to mention Bill Green, a like-minded colleague, equally fascinated by the obscure, overlooked wisdom of the past and its relevance to the future.

Preface

The content for this book developed slowly over many years based on broad, unrelated observations about design, from hand-held objects up to the urban scale. As a long-practicing industrial designer, I have tried to avoid the pitfall of making this book focused on my own métier, in order to reach a broader audience. To that end, I will offer some detailed reflections of designed objects that transcend disciplinary lines and, importantly, provide some necessary background for the more in-depth pages that will follow. Altogether, examples like those which will be shared in the following pages provided the needed foundation for the writing of this book.

As someone who was born and lived in New York City much of his adult life, the perspective in this book is invariably informed by the urban lifestyles of this remarkable city. Famous for embracing the new at the expense of the old, especially in the built environment, I have found myself increasingly curious about earlier, invisible, now demolished phases of the city's development. During the 1990s, when I was in my mid 20s, I would often spend weekends exploring the abundant, diverse neighborhoods across the city's five boroughs. Guided by entries from the *AIA Guide to New York City* and *Forgotten New York*, I rediscovered first-hand New York's earlier built sediments through its skyscrapers, bridges and long-gone demolished infrastructure and technology.[1] One particular Saturday, I came upon an area of West 125th street, near Columbia University, where the street's pot-holed asphalt had worn away, revealing century-old electric streetcar tracks, secured in the ground by Belgian Block pavers. Later, without searching terribly long, I learned the tracks belonged to the 3rd Avenue Railway Co., an expansive *electric* surface transit system that had long since been abandoned in the late 1940s.[2] In a city so congested, often with poor air quality, it struck me very unfortunate that a mostly emission-free[3] surface transit system would be dismantled, in favor of diesel buses which remain the norm.[4] Of course, air-quality and carbon emissions were not much of a concern then as they are now. Today, electric propulsion is booming again, albeit through battery-electric cars, buses, bikes, trucks and other futuristic mobility proposals, including ones that make use of disused urban rail infrastructure.[5] At the same time, many electric rail systems like the 3rd Avenue Railway Co. have survived in other cities and countries, and new systems have been created more recently. But why would so much infrastructure, built so carefully at great expense, be destroyed in the first place? As time passed, I wondered if there were other examples of relevant, seemingly "greener" technologies like this that have also been buried, literally and metaphorically.

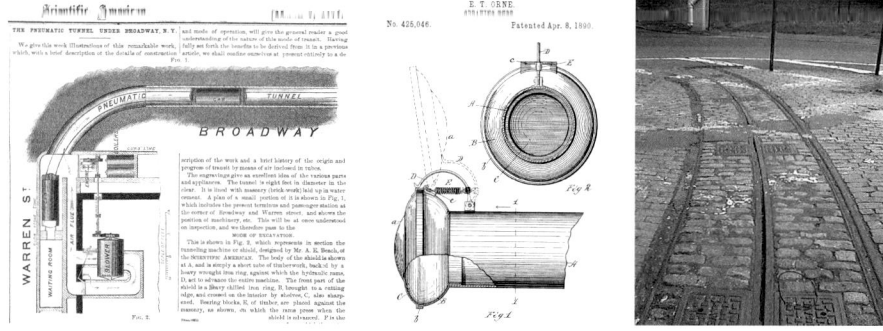

Figure 0.1
Above left: A *Scientific American* article describing Alfred E. Beach's 1870 experimental but ultimately unsuccessful pneumatic subway, in lower Manhattan. Source: New York Times 1870.

Figure 0.2
Above center: An 1890 patent for a speaking tube improvement. These early communication devices were used sporadically before electric intercoms in houses and apartment buildings. Source: USPTO.

Figure 0.3
Above right: A 1995 photograph of abandoned electric streetcar tracks in Manhattan. Source: Brook Kennedy.

Decades afterward, while visiting my wife's family in Switzerland, I later "discovered" (as an ignorant American) traditional passive techniques for cooling buildings that are still used today around the world. Over the summer, like in the US, temperatures in Switzerland can exceed 90F with moderate humidity, yet fewer than 10% of Swiss households have air-conditioning (compared with roughly 90% in the US).[6,7] Certainly, there are many parts of southern American latitudes that are consistently hot and humid, making it difficult to live comfortably or be productive without artificial cooling. This has been especially true recently as unprecedented heat waves have swept through Europe, North America and South Asia in 2021 and 2022.[8] In these extremes, heat can pose health risks to vulnerable age groups like the elderly,[9] and studies have now even linked mechanical cooling to declining temperature-related mortality.[10] But often modern buildings in the US are designed to be sealed shut so that you can *only* use air conditioning, even in moderately warm weather. Opening a window, even on a cooler morning, isn't even possible, and often the AC keeps running day and night. Sometimes window-unit ACs keep windows from being opened at all or windows are painted closed, sometimes for security reasons. In other cases, air temperature is regulated by automated building-wide climate control regardless of the outside weather conditions, or variations in occupants' definition of thermal comfort.[11] Sealed, and often excessively air-conditioned, spaces have increasingly become the norm, carrying tremendous energy and environmental cost.[12] In the near future, many warmer countries are projected to significantly increase their use of AC, and some have struggled to provide enough electric capacity to meet demand.[13]

In contrast to the widespread use of AC in countries like the US, reliance on traditional, passive and low-energy cooling techniques lives on in much of the rest of the world. For example, the strategic practice of opening and closing windows and shutters in the morning, when air temperatures are cooler, remains common. In many countries, by mid-morning before the sun's heat intensifies, awnings are drawn, windows are shut and shutters are closed to help seal cooler air indoors. In some cases, window shutters are closed on the sides of houses with morning sun exposure. By the end of the day in mid-July, house interiors can remain roughly 15 degrees Fahrenheit or cooler than outside. At first, I believed that the superior insulation of thick masonry walls in my in-laws 1910 house enabled this passive cooling, yet I was able to achieve respectable results in my 1946 wood-framed house, in a rural US town (8 degrees cooler in informal tests, observed with a network of connected thermometers). Granted, I also periodically used a small dehumidifier during high humidity weather, but this technology uses considerably less energy than a window unit AC. Even so, after performing this morning window opening process, visiting afternoon guests would often comment how nice the "AC" felt once inside. Given that AC accounts for roughly 12–17% of residential energy use in the US, it was impressive to experience first-hand how these passive cooling practices, or "technologies," namely, methodically opening functioning windows and using shutters, could cool a home reasonably well resulting thereby in significant energy savings. Although it has not been scientifically measured for this book, these savings might rival those of smart thermostats which can pre-emptively start the AC before a home owner's arrival.[14,15,16] While this realization is, of course, not newsworthy to perhaps millions of residents around the world, in wealthier countries that have gotten used to the comforts of relying solely on AC, these kinds of technologies and practices offer relevant alternatives in spite of their relative obscurity. Strangely, there are so many examples of houses built in the US prior to the Second World War that were designed with these principles but whose features are no longer used or understood. Shutters and awnings in the US are now mostly decorative appliqué.

On growing university campuses where development forges ahead, some older buildings are naturally being demolished and replaced with newer larger ones. One example at my own university, the University Club, was built in 1929 as a community gathering space. The architecture, while handsome (if unremarkable in its neo-Georgian style), employed several common passive cooling techniques, including openable, screened, floor to ceiling windows on either side of the first floor for promoting cross ventilation. Inside, a central staircase was designed to help draw hotter air up and out of the building through a roof vent. In the years just prior to its demolition, other newer buildings would all but certainly have had the air conditioning on around the clock, even on milder days. When the university expanded to promote new initiatives with innovation, the University Club was chosen as a site for a new center. Without pause, a large new building was constructed in its place. Unfortunately, the new building was designed without any of the passive cooling techniques of its predecessor, choosing instead to have AC alone cool the interior. Ironically, the energy hungry new building, without operable windows or passive cooling options, was crowned a center for innovation on campus. The University Club with it near-invisible alternative cooling technology was sacrificed as an obsolete relic.

Of course, so many new innovations provide valuable human benefits. This is partially why their promise is compelling and their stories are so powerful. Without question, countless recent technological innovations have enhanced the human condition and brought immeasurable value: solar technology has advanced renewable energy by harnessing the immense energy potential of the sun (even if old photovoltaic cells cannot yet be rehabilitated or recycled). Vaccines have saved countless lives during the Covid-19 pandemic. Medical treatments have mitigated the effects of cancer, and other terminal ailments and illnesses. Domestic unmanned aerial vehicles (UAVs) can expedite the delivery of blood transfusions and transport critical medicines to hard-to-reach areas. Robotic prosthetics offer the possibility to restore limbs to amputees and wounded veterans. In writing this book, the auto spelling checker has helped save time. The list, of course, goes on into every conceivable area of human experience. But just as new technology offers hope to improve the human condition and remedy environmental degradation, there are also alternative passive and traditional technologies that could seemingly *also* be useful again. Much of this useful knowledge is being forgotten. This book will explore examples across the built environment to raise awareness of these options while encouraging designers to consider their adoption as they envision a more ideal, resource efficient future in the century ahead.

Certainly, there are also examples of now abandoned "breakthrough" technologies that should remain so, especially in areas affecting human health. Luminous radium (glow-in-the-dark) paint is one such example that has now been banned from the marketplace.[17] It took decades for health officials to conclude that substances like these, no matter how useful, posed health risks to humans and the environment. Clean-up of radioactive contaminated soil used from the US Radium Corporation's luminous paint factory ultimately cost taxpayers millions of dollars in environmental remediation through the US EPA's Superfund program.[18] Today, many other substances have been or are currently under review by health agencies, notably Asbestos and Bisphenol A, the latter which is found in plastics and resins used in baby bottles and food packaging.[19] Some of these substances will likely be banned as well. But for now, let us refocus our attention on future possibility by exploring examples of alternative technologies across the built environment that offer potential value looking ahead.

Notes

1 Dumas, A. (1896) 'Alfred Ely Beach,' *Scientific American*, 74(2), January 11 [online]. Available at: www.jstor.org/stable/26119718?seq=1#metadata_info_tab_contents (Accessed: July 29, 2022).

2 Ballard, C.L. (2005) *Metropolitan New York's Third Avenue Railway System*. Cambridge: Arcadia Books.

3 Emission-free refers to the vehicle itself. Often fossil fuels are used to produce the electricity that powers electric vehicles, even today. US Energy Information Agency (2022) *Electricity explained*, July 15 [online]. Available at: www.eia.gov/energyexplained/electricity/electricity-in-the-us.php (Accessed: July 28, 2022).

4 While diesel and natural gas buses remain dominant, battery-electric buses are slowly replacing them in public transit fleets around the world. World Bank (2021) *Electrification of public transport: A case study of the Shenzhen Bus Group. Mobility and Transport Connectivity* [online]. Available at: https://openknowledge.worldbank.org/handle/10986/35935 (Accessed: July 28, 2022).

5 Dezeen Magazine's 2022 Future Mobility Challenge featured numerous electric mobility concepts, one that even proposed repurposing electric streetcar tracks. Amber, A. (2022) *Abacus electric tram proposal repurposes disused railways* [online]. Available at: www.dezeen.com/2022/06/30/abacus-lea-haats-erik-mantz-hansen-konstantin-wolf-future-mobility-competition-arrival-finalist/ (Accessed: July 28, 2022).

6 Randazzo, T., De Cian, E. and Mistry, M.N. (2020) 'Air conditioning and electricity expenditure: The role of climate in temperate countries,' *Economic Modelling*, 90, pp273–287.

7 From 1993–2009, US households with air conditioning rose from 68% to 87%. Today it has surpassed 90%. U.S. Energy Information Administration (2011) *Air conditioning in nearly 100 million U.S. homes*, August 19 [online]. Available at www.eia.gov/consumption/residential/reports/2009/air-conditioning.php (Accessed: February 22, 2022).

8 Khullar, D. (2022) 'Living through India's next-level heat wave,' *The New Yorker*, August 1 [online]. Available at: www.newyorker.com/magazine/2022/08/01/living-through-indias-next-level-heat-wave (Accessed: August 1, 2022).

9 The Environmental Protection Agency (2020) *Climate change indicators: Heat-related deaths (1979–2018)* [online]. Available at: www.epa.gov/climate-indicators/climate-change-indicators-heat-related-deaths (Accessed: July 29, 2022).

10 Barreca, A., Clay, K., Deschênes, O., Greenstone, M. and Shapiro, J.S. (2016) 'Adapting to climate change: The remarkable decline in the U.S. temperature-mortality relationship over the 20th Century,' *Journal of Political Economy*, 124(1), pp105–151.

11 Roaf, S., Nicol, F. and Humphreys, M. (2016) *Adaptive thermal comfort: Foundations and analysis*. Abingdon, Oxon: Routledge.

12 US Energy Information Agency (2018) *Air conditioning accounts for about 12% of U.S. home energy expenditures* [online]. Available at: www.eia.gov/todayinenergy/detail.php?id=36692 (Accessed: July 29, 2022).

13 *The Economist* (2022) 'Indian power plants are running out of coal,' May 7 [online]. Available at: www.economist.com/asia/2022/05/07/indian-power-plants-are-running-out-of-coal (Accessed: August 7, 2022).

14 Insufficient evidence exists to compare passive technology savings with smart thermostats. Some estimates have suggested roughly 10% energy savings while others have shown that they are often misused and in some cases cause home owners to use more energy. Sage, S. (2021) 'How much will a smart thermostat really save on energy costs?' *Digitaltrends* [online]. Available at: www.digitaltrends.com/home/how-much-will-a-smart-thermostat-save-on-energy-costs/ (Accessed: July 29, 2022).

15 McHugh, C.J. (2017) 'Evaluation of the Nest Learning thermostat in a multifamily apartment setting.' University of Georgia. [Online]. Available at: https://esploro.libs.uga.edu/esploro/outputs/graduate/Evaluation-of-the-Nest-Learning-thermostat/9949333451602959 (Accessed: July 29, 2022).

16 Alderman, Z.E. (2017) *Exploring energy, comfort, and building health impacts of deep setback and normal occupancy smart thermostat implementation*. Graduate Theses and Dissertations. Available at: https://scholarworks.uark.edu/etd/2531 (Accessed: July 29, 2022).

17 Moore, K. (2017) *The Radium Girls: The dark story of America's shining women*. Napierville: Sourcebooks.

18 The Environmental Protection Agency (2005) *Glen Ridge, New Jersey's EPA superfund site* [online]. Available at: https://cumulis.epa.gov/supercpad/cursites/csitinfo.cfm?id=0200996 (Accessed: July 29, 2022).

19 Borrell, B. (2012) 'US opts not to ban BPA in canned foods,' *Nature*, April 1 [online]. Available at: https://doi.org/10.1038/nature.2012.10370 (Accessed: July 29, 2022).

Introduction

Alternative Technology for the 21st Century

This book explores synergies across two seemingly unrelated topics: alternative and traditional technology in design and architecture, and global efforts to address broad environmental problems. In summation, it asks the following simple question: can alternative, traditional and often abandoned technology be useful again to help address present environmental challenges? Most likely, to suggest this may seem unrealistic or nostalgic as abandoned technologies should perhaps be forgotten. Notwithstanding, examples will be presented that challenge the assumption that what is new is inherently superior, and even suggest that many known and largely forgotten alternative technologies deserve to be reexamined, and possibly reintroduced alongside, or in place of, today's newest innovations. Examples will be broad, and will include passive, manual and forgotten energy efficient technology, durable design approaches and traditional renewable materials that are being rediscovered. Ultimately, this book aspires to raise awareness of these technologies, some better known and others more obscure, to encourage designers to make their use *desirable* again for today's milieu.[1] What industrialized society might have emphasized in past years—human-centeredness, convenience and comfort come to mind—are no longer exclusively prioritized now.[2] Instead, climate change, pollution and resource conservation are top of mind.[3] At the same time, society runs short on ideas to best address these ongoing concerns. Hopefully this book provides a fresh, additional perspective in service of these efforts.

The content of this book develops through the following topics. In Chapter 1, a theoretical basis is established via an exploration of three foundational themes: an understanding of the United Nation's Sustainable Development Goals and current progress towards achieving them; a more detailed discussion and definition of the term *alternative technology*; and finally, an examination of the ideas leading up to the present cultural commitment to, and faith in, in *new technology*. In Chapter 2, we will begin looking at examples of alternative technology through the evolution of the city street from pedestrians, bicycles and electric streetcars to automobiles and buses—and in a few cases their return. Following this chapter, Chapter 3 explores alternative methods of cooling and heating buildings, some hundreds of years old. Chapter 4 examines natural daylighting and its growing relevance today for energy conservation and health. Chapter 5 surveys cleaning and maintenance in the home. Chapter 6 covers an overall encouraging trend in product design to promote more durable, less disposable design, similar to practices that were once commonplace less than one hundred year ago. Chapter 7 explores traditional, renewable materials that are relevant for maintaining natural resources for a growing human population. Chapter 8 covers alternatives to electrifying some appliances and tools in the home. Chapter 9 speaks about manufacturing goods with renewable energy and Chapter 10 finishes off by reexamining past low-energy transportation technologies that

DOI: 10.4324/9780367814304-1

could be reinvented for today. Finally, in the Conclusion, the book summarizes emerging pat
terns found in alternative technologies, to offer relevant, new perspectives for design in the
21st century. Together they follow four overlapping traits: alternative technologies use *passive*
or *harvested energy* from natural sources; some are *manually* powered or *hybridized* with
manual override to offer opportunities to engage in physical activity while conserving energy.
Most can be easily repaired and use renewable materials to support a Circular Economy.
Generally, they all derive from a spirit of openness and respect for alternative, traditional practices and knowledge with an orientation towards resource conservation.

Audience, Scope and Focus

Exploring alternative technology is intended to offer fresh perspectives for designers,
but it is also intended to reach an academic and public audience. It is hopefully accessible to read by anyone interested in the topic. It is by no means an historical volume or a
work of sociology or economic theory. Most of all, the book reflects the point of view of
a designer, with the intention of inspiring creative professionals to *reimagine alternative
technology for sustainable design.* While human-centered design (Design Thinking) provides value to users and businesses, and is widely practiced by the private sector within
a customer-focused system,[4] this book focuses more on longer-term environmental
questions and concerns. Others have written on related topics before, many in the
design professions, like Jonathan Owen, who has recently questioned the reliance on
consumer behavior to change patterns of resource consumption and other societal ills.[5]
The widely influential Victor Papanek has envisioned a world where design serves as
a force of responsibility by helping reduce overconsumption, pollution, starvation and
other global challenges.[6] Additional voices have pushed this topic further: academic
Terry Irwin and collaborators have even proposed an evolved design field, defined as
Transition Design, which relies on broader Socio-Technical systems to bring about fun-
damental changes in consumer behavior.[7] Finally, economist Ernst Schumacher has
argued for the concept of "enoughness" or the need to balance human needs with
local, appropriate use of technology and sustainable resource management.[8] While this
book acknowledges the need for greater systemic change to occur within the design
professions to prioritize global environmental challenges, broader design systems will
not be the principal focus. Instead, the book offers a form of *discursive design*, based
on alternative technology, seeking to encourage discussion, and action, to promote
desirable future realities.

Gender, Labor and Motorized Appliances

Many of the examples discussed in the following chapters discuss the energy-conserving
value of reintroducing human power to supplement electric and motorized appliances.
Without a doubt, there has been ample evidence and scholarship focused on the ability of
electric-powered appliances, along with other inventions, to reduce the drudgery of physical chores—this is especially poignant for the economically disadvantaged and for women
who continue to perform the majority of housework in family homes.[9] Indeed, a great deal
has been written about the impact of labor-saving inventions on the lives of women.[10,11,12]

By examining alternative technologies that propose some reconsideration of manual effort (bicycling vs. driving), for reasons of energy conservation and human health, this book is by no means suggesting "turning back the clock" on the social advancement of women or any other group. As Ruth Cowan concluded in *More Work for Mother: Ironies of Household Technologies from the Open Hearth to the Microwave*, so many new powered appliances and technologies had a non-intuitive outcome: they only increased expectations of domestic cleanliness rather than reducing overall effort and drudgery.[13] The hope is that alternative technologies can be reintroduced to simultaneously meet the need for reasonable domestic upkeep while reducing waste and energy consumption. An additional aim is that human-powered appliances, like the many examples discussed in Victor Papanek's *Design for the Real World*,[14] could also provide some opportunity for modest, elective physical activity that has been assumed by exercise gyms and moderately rediscovered through upgraded infrastructure in already bikeable and walkable cities.[15]

Western Cultural Bias

The breadth of the examples cited in this book is broad but also incomplete. It is predominantly focused on trans-Atlantic, Western technology and design traditions, although there are some exceptions. This was certainly not the original intention, nor will this study be complete until a more comprehensive, global, intercultural study has been completed. The reason for this unintended outcome was strongly influenced by Covid-19-related travel restrictions and closures of cultural institutions between 2020–2022 when the book was being developed. To be sure, there are numerous historical resources around the world to research alternative technologies of the world's many and diverse cultures. The Smithsonian Institution's Lemelson Center for the Study of Invention and Innovation in Washington D.C. remained closed during much of this time as was the American Museum of the American Indian. Fortunately, online resources came partially to the rescue as access to the Library of Congress and trade publications such as *Sweets Catalogue*, *The American Architect and Building News* and the *USPTO* remained available. All provided invaluable material for exploring alternative technologies in Western culture. Additionally, access to the writings of journalist Kris De Decker's *Low-Tech* magazine, which continues to examine a breadth of alternative and past technologies, has been abundantly helpful and insightful. Nevertheless, this book makes no claim that the examples covered in the pages to follow represent a complete survey of the subject matter. Rather, this book freely acknowledges this limitation, hoping that others will expand the breadth of this topic and pursue similar research projects of their own. Fortunately, some authors already have, and recently. In particular, Julia Watson's *Lo-TEK: Design by Radical Indigenism* covers global sustainable technologies and practices that have endured for centuries for their ingenuity and suitability to local environments.[16] Many other examples exist that are difficult to credit in their entirety. Industrial Designer Brian Skeet, for example, an Anastasi Indigenous American from what is now the US state of Arizona, spoke candidly about the vital importance of reversing the colonial erasure of indigenous inventions in the article *Indigenizing Industrial Design*.[17] The example of the Havusupai Boiling Basket in Chapter 4 is an important example of such a forgotten technology. All design and technology traditions which have been overlooked or pushed to extinction deserve to be examined thoroughly and preserved as a part of this book's larger research aims. Indeed, much more work needs to be done to culturally diversify this topic and to more comprehensively address the promise, and future relevance, of alternative technology.

At the same time, some emphasis on the United States is warranted. The US has eagerly, and somewhat uniquely, embraced new technologies, sometimes at the expense of more energy-efficient alternatives (passive cooling techniques, electric surface rail transit, etc.). In contrast, industrialized nations on other continents, in attempts to uphold environmental standards and preserve their cultural traditions, have continued to use some of the alternative technologies discussed in this book. One of the most prominent examples of a particularly American phenomenon has been the early adoption of, and now *dependence* on, the personal automobile. While the automobile has provided tremendous value—in access to jobs, economic growth and independence[18]—so many American communities now rely on automobiles exclusively, in absence of other viable options, like public transportation. Cities and landscapes have in many ways irreversibly changed to suit automobile lifestyles, with ongoing environmental consequences.[19] Even with the advent of electric vehicles, public transit remains a more energy-efficient strategy next to the personal automobile.[20,21,22] Unfortunately, many developing countries often emulate America's car-centric transport behaviors and urbanism.[23,24] Hopefully, discussions of this kind in this book might ease the uptake of these largely American behaviors elsewhere as growing economies understandably aspire to similar standards of comfort and convenience.

Notes

1 Echoing the spirit of Vitruvius in *De Architectura*, that all buildings must have three attributes: *Firmitas*, *Utilitas* and *Venustas* ("strength," "utility" and "beauty"), design must ensure that alternative technologies are not just utilitarian and efficient—they must be presented beautifully, with *Venustas* to be accepted today.

2 Life-Centered Design is an emerging design framework based on Human-Centered Design (HCD) which has been widely influential in product design practice over the past 20–30 years. Life-Centered Design differs from HCD in that it values all life, not just human need at an ecosystem level.

3 Following the invasion of Ukraine in early 2022, gas prices spiked creating concerns about the cost of heating in European markets that depended on these national resources.

4 Brown, T. (2008) 'Design thinking: Thinking like a designer can transform the way you develop products, services, processes—and even strategy,' *Harvard Business Review* [online]. Available at: https://hbr.org/2008/06/design-thinking (Accessed: August 13, 2020).

5 Owens, J. (2019) '10 principles of Life Centered Design,' June 3 [online]. Available at: https://medium.com/the-sentient-files/10-principles-of-life-centered-design-3c5f543414f3 (Accessed: August 1, 2022).

6 Papanek, V. (2019) *Design for the real world*. 3rd edn. London: Thames and Hudson.

7 Irwin, T. (2015) 'Transition design: A proposal for a new area of design practice, study, and research,' *Design and Culture*, 7(2), pp229–246. DOI: 10.1080/17547075.2015.1051829

8 Schumacher, E.F. (1999) *Small is beautiful: Economics as if people mattered: 25 years later*. Vancouver: Hartley & Marks.

9 Hess, C., Ahmed, T. and Hayes, J. (2020) 'Providing unpaid household and care work in the United States: Uncovering,' *Inequality Institute for Women's Policy Research* [online]. Available at: https://iwpr.org/wp-content/uploads/2020/01/IWPR-Providing-Unpaid-Household-and-Care-Work-in-the-United-States-Uncovering-Inequality.pdf

10 Strasser, S. (1982) *Never done: A history of American housework*. New York: Pantheon Books.

11 Hardyment, C. (1988) *From mangle to microwave: The mechanization of household work*. Cambridge: Basil Blackwell.

12 Yarwood, D. (1983) *Five hundred years of technology in the home*. London: B.T. Batsford.

13 Cowan, R.S. (1985) *More work for mother: The ironies of household technology from the open hearth to the microwave*. New York: Basic Books.

14 Papanek (2019).

15 Bernhard, A. (2020) 'The great bicycle boom of 2020,' *BBC* [online]. Available at: www.bbc.com/future/bespoke/made-on-earth/the-great-bicycle-boom-of-2020.html (Accessed: August 1, 2022).

16 Watson, J. (2019) *Lo-TEK. Design by Radical Indigenism*. Cologne: Taschen.

17 Skeet, B. (2021) 'Indigenizing industrial design,' *World Design Organization*, June 23 [online]. Available at: https://wdo.org/indigenizing-industrial-design/ (Accessed: August 1, 2022).

18 Baum, C.L. (2009) 'The effects of vehicle ownership on employment,' *Journal of Urban Economics*, 66(3), pp151–163. DOI: 10.1016/j.jue.2009.06.003

19 Union of Concerned Scientists (2014) 'Car emissions and global warming,' July 14 [online]. Available at: www.ucsusa.org/resources/car-emissions-global-warming (Accessed: August 1, 2022).

20 Mead, L. (2021) 'The road to sustainable transport. Still only one Earth: Lessons from 50 years of UN Sustainable Development Policy,' *International Institute for Sustainable Development*, May 24 [online]. Available at: www.iisd.org/articles/deep-dive/road-sustainable-transport (Accessed: August 14, 2022).

21 Manjoo, F. (2021) 'There's one big problem with electric cars,' *The New York Times*, February 8 [online]. Available at: https://www.nytimes.com/2021/02/18/opinion/electric-cars-SUV.html (Accessed: August 14, 2022).

22 Harrabin, R. (2019) 'Electric cars "will not solve transport problem," report warns,' *BBC*, July 5 [online]. Available at: www.bbc.com/news/uk-48875361 (Accessed: August 14, 2022).

23 Li, Y., Miao, L., Chen, Y. and Hu, Y. (2019) 'Exploration of sustainable urban transportation development in China through the forecast of private vehicle ownership,' *Sustainability*, 11(16), 4259. DOI: 10.3390/su11164259

24 Ma, L., Manhua, W., Tian., X., Zheng, G., Du., Q. and Wu., T. (2019) 'China's provincial vehicle ownership forecast and analysis of the causes influencing the trend,' *Sustainability*, 11(14), 3928. DOI:10.3390/su11143928

Chapter 1

Why Should Alternative Technology Be Reexamined?

Slow Progress Towards the United Nations Sustainable Development Goals

Based on studies published by various international agencies, considerable difficulties confront life on earth in the 21st century: growing populations and diminishing resources,[1] globalization and aspirations for improved quality of life, loss of biodiversity and pollution, rising sea levels and climate change, and conflict and inequity related to all of the above. The United Nations (UN), established the Sustainable Development Goals to enact new measures and promote fresh ideas to eradicate these seemingly intractable challenges.[2] The 2015 Paris Climate Treaty, for instance, is one such measure creating binding commitments to limit global warming.[3] To date it has been signed by nearly 200 countries. However, as of 2020, the UN conceded that many of the Sustainable Development Goals are not on target, although certainly not for a lack of effort. Indeed, there have been abundant initiatives and proposals, many by designers and inventors, to help achieve or surpass the UN's targets.[4] For all of these many well-intentioned new ideas, often driven by the promise of emerging technology (and published yearly in design innovation competitions), these efforts are evidently not enough.[5]

New renewable energy technology and material science offers a glimmer of hope that today's resource consumption behaviors can be perpetuated, be it by wind power, solar power, electric cars or single-use biodegradable water bottles (that don't collect in the world's oceans). Unfortunately, this picture is also incomplete. Environmental studies researcher Richard York has argued that renewable energy has only added to fossil fuel consumption rather than replacing it. In *Do Alternative Energy Sources Displace Fossil Fuels?*[6] York demonstrates this phenomenon through analysis of renewable energy from 1960–2009, showing how renewable energy merely complements energy consumption rather than replacing their non-renewable, carbon-producing alternatives including petroleum, coal and natural gas. He describes renewables as energy "additions" rather than energy "transitions," challenging the popular notion relating to the global adoption of these energy technologies.[7] At present, roughly 20% of US electricity generation derives from renewables compared with roughly 50% in Denmark, a recognized world leader in renewable wind-turbine technology.[8] Recycling of plastics, metals, timber and more offers some hope, just like renewable energy. However, despite the slowly increasing rates of recycling globally, use of these raw materials is also growing, meaning that York's energy analysis likely applies to material consumption as well.[9]

Demand and consumption of water is also increasing in the home and for industrial use.[10] Technologies like desalination, which uses energy to remove salt from sea water to make it drinkable is being adopted in regions lacking alternatives but otherwise fortunate

 DOI: 10.4324/9780367814304-2

to have the energy resources to use it. In 2018, South Africa approached "Day Zero," or the point where tap water had been completely exhausted. Amazingly, the country managed to avoid this calamity, not with technology but through education, awareness and conservation efforts.[11] Meanwhile, the American West is facing a similar conundrum with a multiyear drought, evident in the falling levels of the Colorado River which supplies much of the south west US. The City of Las Vegas has managed to reuse significantly more of its water by tightly regulating water use.[12] An important lesson from these examples is that *conservation* and techniques like Behavior Change remain equally important as technological innovation to manage resource consumption. In this book, examples of alternative and traditional technology can provide further insight for managing resources, like water, by observing how they were managed before they were as readily accessible as they are today.

Conservation Culture: Wartime Austerity Programs of the Second World War

The Second World War marked one of the most horrific chapters in recent human history, one that has been written about extensively. During the war, conservation played a vital role to support wartime manufacturing efforts. The US Office of Price Administration (OPA), for instance, implemented a rationing program to limit consumption of precious resources: gasoline, metals, food and other commodities, which were metered out in limited supply. Metals of all kinds were recycled for use in the manufacture of airplanes, tanks, ammunition and more.

Figure 1.1
A Second World War US Bureau of Home Economics poster promoting resource conservation. Other advertisements of the period encouraged saving metals and other household materials that could be used for making weapons to help fight the war. A wartime culture of conservation was valuable for addressing challenges of the time. Cultivating a similar conservation culture could serve similar purposes today. Source: US Bureau of Home Economics.

Today, these important wartime programs, and the spirit of conservation they inspired, are easily overlooked and forgotten. The mere thought of such programs is now unthinkable within today's customer-focused expectations. After the war, as these OPA programs ended, conservation culture quickly transformed to a thriving consumer economy in the decade that followed. Federal housing loans and subsidies helped some returning veterans go to college and buy homes. At the time, the average size of newly built family houses was around 1000sq ft. Since 1950, average house sizes have more than doubled.[13,14] At first, post-war houses were simpler—lacking the typical technological comforts taken for granted today. Over time, as households were accessorized with air conditioning, more electric appliances, televisions, multiple cars and other accoutrements, their per-capita consumption of energy, water and other resources increased as well.[15] Eventually, many of these amenities became standard equipment; suburban housing developments became larger and more spread out, sidewalks and walking fell out of favor and cars began to overtake public transportation. Electric streetcars and interurban trains were dismantled especially following the Second World War. Today, American society lives with the consequences of these decisions as walkability is valued again.[16] It is hard to believe that more walkable communities were once the norm not so long ago. Today, walkable suburban towns in the eastern US, with access to older commuter transportation networks, command higher real estate prices.[17,18] In a short period of time since the Second World War, consumer habits and energy consumption has changed dramatically in the US and other countries. In the following chapters, this book will examine some of these alternative technologies used in homes and communities in the post-Second World War period, many that could be particularly relevant again today.

Defining *Alternative Technology*

In recent years, if asked to identify the most commonly used catch phrases in design and business language, "disruptive technology" would surely rank among them.[19] Alongside other concepts, like "innovation" and the "serial entrepreneur," disruptive technology has become international *lingua franca*, describing any remotely new idea, in design, engineering, architecture, education and most any other field.[20] Silicon Valley, associated with risk-taking technology startups and innovation culture, has done much to define these terms. Design firms in the region have only further supported this cultural and economic phenomenon. In *Make it New: A History of Silicon Valley Design*, author Barry Katz explains how Design Thinking techniques only reinforced Silicon Valley's authority as the capital of disruptive innovation, by translating new technologies into products and services that people understand and actually want.[21] There and elsewhere, billions of dollars in venture capital have given birth to countless examples of disruption: the personal computer, smartphones, artificial intelligence and autonomous cars, among many others. Many of the resulting commercial organizations formed from these technologies include a dizzying number of startups and some of the largest and most powerful companies in the world. Alphabet, Tesla, Apple, Meta and so many others have seen their market values rise to historically unprecedented levels in recent years. On top of financial success, the rhetoric of the disruptive technology economy often promises utopian visions.[22] Altogether, Silicon Valley is admired worldwide and inspires billions of dollars in investments to emulate elsewhere.[23,24]

Perhaps the most celebrated example of a disruptive technology today is the ubiquitous smartphone. Estimates vary, but more than seven billion of these devices are in use today. Since the introduction of the Apple iPhone in 2007, human behaviors and routines have measurably changed. Smartphones can adjust home thermostats remotely so the AC is on when people get home. They can summon ride hailing services, connect us socially in novel ways or simply be used to call Mom. At the end of a busy work day, they can stream soothing sounds to put people asleep, offer reminders to walk more steps and check heart rates. The variety and depth of services is immense and grows tremendously every year. At the same time, the impacts of these devices are numerous and well-documented. Millions of smartphones are discarded every year contributing to significant and growing volumes of hazardous electronic waste.[25,26] In addition, their overuse has been linked to sedentary behaviors, eroded social skills and depression.[27] Further, despite the sophistication and power of technologies like these, life expectancy, personal and job satisfaction has remained unchanged or has declined.[28] Many are now choosing to *unplug* these devices seeking a healthy balance with always-on mobile connectivity.[29] Additional examples of prominent disruptive technologies are numerous; automobiles, for example, have offered freedom, job access and made life more convenient, but have also increased per-capita energy use, pollution and discouraged walking and other physical activity. Air conditioning, as mentioned before, has provided thermal comfort especially during the latest extreme heat events—at the same time it represents one of the highest energy growth areas.[30] And yet, since the advent of these innovations, the optimistic rhetoric of disruptive technology remains pervasive: life-changing, transformative and inevitable.

Increasingly, social critics have challenged the assumption that disruptive technology has measurably improved quality of life for everybody. Writer Alison Arieff, for instance, noted in her 2016 *New York Times* editorial "Solving all the Wrong Problems":

> We are overloaded daily with new discoveries, patents and inventions all promising a better life, but that better life has not been forthcoming for most. . . . [I]n San Francisco, where such companies are based, sea level rise is ominous, the income gap between rich and poor has been growing faster than in any other city in the nation.[31]

Much in the spirit of Arieff's perspective, the focus of this book will be to help solve *relevant problems*, especially those in alignment with the UN's Sustainable Development Goals.

Disruptive Technology vs. Relevant, Forgotten Precedents

If asked, most would likely consider electric cars a form of disruptive technology (even though they were first invented in the 19th century). Conversely, others might not regard bicycles (unless electrified) as "technologies," but here in this book they will be defined as such—specifically as *alternative technologies*. Alternative technology can fall into many categories: abandoned, forgotten, underused, traditional and pre-industrial. Unlike cars on a road, bicycles and even sidewalks are defined here as alternative technologies: human-contrived means of getting from one place to another. As mundane and unsexy as they might seem, a bicycle, or a smooth, level sidewalk for walking, are equally types of mobility technology. For the able bodied, cars (electric or otherwise) cannot compete with bicycling or walking short distances in terms of saving energy, costs, reducing emissions and promoting health benefits. Of course, not all commutes are short—some cannot be controlled and homes that are closer to walkable city centers are generally less affordable and relatively uncommon in spread out countries like the US. Additionally, not everyone is able to walk. Still, next to cars on trafficked roads, the traditional sidewalk continues to be a relevant technology today and some cities are doing more to expand them. Nevertheless, experts insist much more investment is needed.[32] They can be designed to be more spacious, to allow access to wheel chairs and the differently abled, to have abundant tree shade and offer adequate signage and provisions for crossing streets safely. Based on their original purpose, to allow people to walk comfortably, their design has advanced tremendously in the past two centuries to remain relevant today and in the future. In the past 20 years, some cities, including New York, have rebuilt their pedestrian spaces, some reclaiming territory from streets which were widened in the mid 20th century to accommodate more automotive traffic.[33] Further sidewalk expansion has been fueled by the Covid-19 pandemic to promote social distancing. In the following chapters, similar examples to the sidewalk, many less visible, will be examined. These sections will consider how these *alternative technologies* could be reimagined; again, to support the UN Sustainable Development Goals and promote human health.

Origins of Technological Progress: How Did New Become *Better*?

"Under the idea of mechanical progress, only the present counted, and continual change was needed in order to prevent the present from becoming passé, and thus unfashionable. Progress was accordingly measured by novelty, constant change and mechanical difference, not by continuity and human improvement."[34]
—Lewis Mumford, 1962, in *The Case Against Modern Architecture*

"Industry pandered to the public's ready acceptance of anything new, anything different . . . . and artificially accelerated consumer whims gave birth to the dark twins of styling and obsolescence."[35]
—Victor Papanek, 1971, in *Design for the Real World*

As Lewis Mumford and Victor Papanek suggest in their own ways, novelty or newness has a powerful effect on culture and society. The new is different, surprising, even entertaining; it garners attention and makes what preceded it seem boring or outdated. The hypnotic power of newness, and by extension *technological progress*, remains seductive, yet the idea is based on assumptions that have been developed and nurtured slowly over centuries, rooted in philosophy, science and other disciplines. In this section, the origins of the idea of "progress" will be broadly summarized, to understand better, if imperfectly, how the ongoing fascination with novel technology came into being.

The present-day conception of technological progress owes many of its roots to the Enlightenment,[36] including the philosophical writings of Auguste Comte, Herbert Spencer and other well-known thinkers of the period. Additional precedents can be traced to

19th-century anthropological thought, specifically the concept of *cultural evolution*, derived from Charles Darwin's theories of *biological evolution*.[37] Additional, broader ties to classical philosophy and religion can be made as well. Generally, the premise or construct of technological progress holds that social advancement can be driven by scientific discovery and technical improvement. These ideas are further connected to the well-known work of economist Thorstein Veblen, who is credited with introducing *Technological Determinism*, a theory that technology drives culture not the other way around.[38] Subsequent 20th-century economists have echoed these ideas of progress, notably the widely influential Joseph Schumpeter, who coined the term "Creative Destruction." In *Capitalism, Socialism and Democracy*, Schumpeter suggested that entire industries, often rooted in technology, are dismantled and replaced by newer ones, based on technological advancements.[39] Despite short term losses in jobs, the net gain of techno-industrial Creative Destruction is ultimately beneficial to economies and society. Examples include the printing press, the combine harvester, railroads and the automobile—more recent examples would certainly include the personal computer, the internet and the smartphone. By the 1980s, sociologist Robert Nisbet wrote about the continued influence of these ideas, noting that dominant intellectual discourse continued to pit "progressivists" against "declinists" or those who embrace technological change versus those who obstruct it.[40] In the 1990s, economist Clayton Christensen famously introduced the concept of "disruptive innovation" in *The Innovator's Dilemma* which cemented these ideas into modern language (discussed earlier) and which continues today.[41] Academic theory and popular media frequently reinforce these ideas: the value and the inevitability of disruptive technological progress. Silicon Valley leaders Steven Jobs, Elon Musk, Larry Page and Sergei Brin are celebrated globally for their *Creative Destruction* of presumably inferior industries and ways of life. Again, underlying this belief system is the assumption that technologies mature progressively, simultaneously advancing civilization. Silicon Valley's technology culture has certainly perpetuated this narrative, often promising that new technology (notably "connectivity") will make life easier while addressing environmental ills and political unrest.[42]

While much of this thinking persists today in popular culture, other theories, notably the *Social Construction of Technology*, assert that technological progress itself is a product of cultural values.[43] Technology historian Thomas Hughes, who is closely associated with this theory, wrote extensively about the larger societal and commercial forces that enabled inventorship to flourish on a systems level in *American Genesis: A Century of Invention and Technological Enthusiasm, 1870–1970*.[44] Commercial interests are also frequently cited for undermining competing technologies for economic gain, demonstrated recently by antitrust suits against Alphabet and other large technology firms.[45] Decades ago, the alleged monopolist replacement of popular street car systems in favor of technically inferior buses has also been widely referenced in academic and popular media circles. Further studies have suggested that economics and social constructs of American individualism, not technology, led to the electric streetcar's demise.[46,47,48]

On a parallel historical trajectory, outright critique of technological progress follows a similar path. In the late 18th and early 19th centuries, Luddites, among other groups, decried new (disruptive) manufacturing technologies fearing their negative impact on the livelihoods of laborers, craftspeople and society. As a response, Utopian designer William Morris envisioned a pastoral future society that would be free of industrialization and capitalism.[49] Around the same time, Karl Marx in *Das Capital* famously opposed how technological machinery was used to reinforce inequitable social hierarchies. Roughly a century later, industrial designer Victor Papanek critiqued the industrial design profession for serving commerce and, by extension, the wasteful practice of *planned obsolescence* in *Design for the Real World*.[50] In the 21st century, various technology writers have offered further spirited critique of technological progress. Evgeny Morozov, for instance, in *To Save Everything, Click Here: The Folly of Technological Solutionism*, cautions against the belief that pervasive, big data-driven technologies can solve all of mankind's problems.[51] Others, like researchers Joyce and Michael Huesemann, insist that new technology can neither be introduced without some undesirable, unintended consequence nor save the environment.[52] Finally, Lee Vinsel and Andrew L. Russell's *The Innovation Delusion: How Our Obsession with the New Has Disrupted the Work That Matters Most* poignantly argues how the relentless investment and pursuit of innovation has fallen short of truly advancing society while underfunded and poorly maintained infrastructure crumbles around us.[53]

Progress is also frequently presented as a means of reducing human effort, by alleviating the labor and ennui of daily chores. In the mid 20th century, home cleaning technologies, from vacuums to washing machines, were marketed to families, particularly women, as labor-saving tools, promising greater leisure, comfort and social mobility. Imagery from advertisements and promotional materials of the time often portrayed women happily engaged in housework supported by these labor-saving technologies. But again, as Ruth Cowan has argued, these devices ended up simply raising standards for home cleanliness—not reducing effort. Nevertheless, the dream of labor-saving technology remains a powerful message, from robotic vacuums to home delivery apps. Some evidence, however, suggests that the wind is beginning to shift: the popular Pixar film "Wall-E" offers a satirical view of humanity in the future, where humans are so inundated with conveniences that they simultaneously destroy the environment while languishing as robots cater to their every desire.[54]

Today, in contrast, many examples of comparatively labor-intensive cooking trends have flourished: the rising popularity of traditional bread-baking, roof gardening and composting further suggests that convenience is not the only criteria that drives user behavior. Food that is local, fresh and pesticide free is increasingly valued. Best-selling writer David Sax echoes these sentiments by questioning the human desirability of many significant innovations that were supposed to make life better, contrasting them with "rearward innovations" that are being rediscovered for their deliberate slowness and analog inconvenience, from bikeable streets to brick and mortar libraries.[55] Finally, social trends, including "downscaling" and others, have emphasized living in smaller homes with fewer, more meaningful possessions and eschewing technology for reasons of wellness and environmental conservation.[56] In the notable industrial design research project *Furnishing Utopia* led by John and Wonhee Arndt, an international collective of designers harness traditional, more labor-intensive Shaker methods of craft, joinery and production to reflect upon and advance durable product design and conscious consumption. In their words, "[Furnishing Utopia] is a global collective, exploring how design values are interconnected across cultures and time. By understanding the past, and engaging with the present, we imagine more perfect ways of living for the future."[57]

Figure 1.2
Furnishing Utopia's Shaker-inspired collection of home furnishings are an embodiment of "slow production" or more traditional ways of manufacture whose goal is to promote durability in design and conscious consumption, to contrast with mass disposable norms. Source: Furnishing Utopia.

Technology Introduction as Entertainment

In addition to the venerated role played by new technology in industrial societies, it also provides a form of human amusement and entertainment. For over a hundred years, technology introductions have been promoted, celebrated and fetishized. During the 1854 Crystal Palace Exhibition in New York City, inventor Elijah Otis exhibited the first commercial elevator to a public audience. As incredulous crowds gathered, Otis had the elevator's cable severed to demonstrate the safety of his invention. Spectacles like these became common forms of showmanship. In the mid 20th century, General Motor's 1964 "Futurama" exhibit at the New York World's Fair dazzled audiences with future visions of automotive utopia with cars that resembled science-fiction spacecraft. Auto shows now are much the same. In the 2000s, Stephen Job's introduction of Apple products drew similar crowds of spellbound amazement. These kinds of spectacles are not limited to the product design; they have also extended to the architectural scale. In *Delirious New York*, Architect Rem Koolhaas describes how in the race to build the tallest skyscraper, the secretive introduction of the Chrysler building's spire design echoed earlier amusement park bravado in Coney Island.[58] Today, pavilion exhibits at the recent Expo 2020 Dubai and device introductions on the floors of the Consumer Electronics Show (CES) continue to promote optimistic future visions from the latest technologies. While most that are showcased never succeed in the marketplace, much of the hype is warranted; in the years of this book's writing, techno-impresario Elon Musk, chief of Tesla, has seen the company's valuation rise several-fold as electric cars have found millions of customers worldwide.

Figure 1.3
Above left: Elijah Otis at the Exhibition of the Industry of All Nations personally demonstrating the efficacy of his "Safety Elevator" after his partner severed the main hoisting rope. Source: SSNEG (original author unknown).

Figure 1.4
Above right: Apple co-founder Stephen Jobs famously impressed crowds of journalists and technology enthusiasts at the annual MacWorld exposition showcasing product introductions. Source: Matthew Yohe (CC BY 3.0).

In other cases, the broader implications of new technology are not always obvious, but develop notwithstanding. Ballistics, for instance, advanced impressively from the early 20th century to now and might be described in these terms. Beginning with propulsion experiments by Robert Goddard in the 1920s, rocketry was adapted for warfare shortly afterward by the Nazis, under Dr. Werner Von Braun.[59] Von Braun led the development of the notorious V2 missiles that terrorized the Allies during the Second World War. Afterwards, Von Braun was invited to the United States to adapt the technology for nuclear weapons and, eventually, NASA's space program. Perhaps the most ambitious technological spectacle in human history, the Apollo lunar missions provided an unprecedented moment of reflection about life on Earth. "One small step for man. One giant step for Mankind" proclaimed astronaut Neil Armstrong as he touched down on the surface of the Moon. Today, unmanned rockets launch scientific satellites to explore the solar system and to support telecommunications on Earth. Recently, in 2021, the internationally financed James Webb Space Telescope was launched to provide never-before seen images of our galaxy. Hundreds of billions of dollars later, SpaceX and competitors now vie for commercial domination, while delighting technophiles with live launch and land video feeds, some ending disastrously. Delivered like professional athletic events, these risky, technological spectacles provide ongoing entertainment, pitting human aspiration against the laws of nature. Indeed, tremendous scientific knowledge is often gained; at the same time, innovation itself is promoted, sometimes obscuring the attendant costs.[60]

Many other technologies have developed comparably to those mentioned earlier and comprehensive volumes have summarized their history and value. Among them are Charles Singer's *A History of Technology* and Ian McNeil's *An Encyclopedia of the History of Technology.*[61,62] More popular titles like Steven Johnson's *How We Got to Now: Six*

Innovations That Made the Modern World offer engaging histories of the nuances, consequences and unexpected rise of modern technologies. Speaking generally, popular culture presents new technology heroically and optimistically, and as an inevitable outcome of human evolution. Echoing this trend, universities now promote student and faculty innovation and commercialization as a new part of general education and the academic mission.[63,64] With these precedents in mind, this book will now change its focus to alternative technology. Examples will be examined at all scales of the built environment to explore how they might serve society's goals in the century ahead, alongside the seductive influence of emerging technology.

Figure 1.5
A montage of new technology at a US university with images glorifying augmented reality headsets, unmanned aerial vehicles and a variety of emerging new technologies. Source: Andy Potts, Anna Goodson Illustration Company.

Notes

1 United Nations Environment Programme International Resource Panel (2019) *Global Resources Outlook 2019: Natural Resources for the Future We Want – Summary for Policymakers* [online]. Available at: https://wedocs.unep.org/20.500.11822/27518 (Accessed: August 1, 2022).
2 United Nations Department of Economic and Social Affairs Sustainable Development (2015) *Transforming our World: The 2030 Agenda for Sustainable Development* [online]. Available at: https://sdgs.un.org/sites/default/files/publications/21252030%20Agenda%20for%20Sustainable%20Development%20web.pdf (Accessed: August 1, 2022). This book specifically addresses the following UN SDG's: Clean Water and Sanitation (6), Sustainable Cities and Communities (11), Responsible Consumption and Production (12), Climate Action (13), Life on Land (15) and others.

3 United Nations Climate Change (2015) *The Paris Agreement* [online]. Available at: https://unfccc.int/process-and-meetings/the-paris-agreement/the-paris-agreement (Accessed: August 1, 2022).

4 Recently, advocacy organization *Architects Declare* has outlined a pledge to find ways to reduce the environmental impact of buildings and construction. *Architects Declare* (n.d.) [online]. Available at: www.architectsdeclare.com (Accessed: August 1, 2022).

5 Global innovation competitions promote design-based ingenuity to address global environmental challenges. These include the Holcim Award (www.holcimfoundation.org/awards/6th-cycle), the James Dyson Award (www.jamesdysonaward.org/en-US/), the Index Award (https://theindexproject.org) and many others.

6 York, R. (2012) 'Do alternative energy sources displace fossil fuels?' *Nature Climate Change*, 2, pp441–443. 10.1038/nclimate1451

7 York, R. and Bell, S.E. (2019) 'Energy transitions or additions?: Why a transition from fossil fuels requires more than the growth of renewable energy,' *Energy Research & Social Science*, 51, pp40–43. ISSN 2214–6296. https://doi.org/10.1016/j.erss.2019.01.008

8 US Energy Information Agency (2021) *Electricity explained: Electricity use in the United States* [online]. Available at: www.eia.gov/energyexplained/electricity/electricity-in-the-us.php (Accessed: August 8, 2022).

9 Ritchie, H. and Roser, M. (2022). *Plastic pollution/our world in data* [online]. Available at: https://ourworldindata.org/plastic-pollution (Accessed: August 1, 2022).

10 United Nations Water (2019) *UN World Water Development Report* [online]. Available at: www.unwater.org/publications/world-water-development-report-2019/ (Accessed: August 1, 2022).

11 Onishi, N. and Sengupta, S. (2018) 'Dangerously low on water, Cape Town now faces day zero,' *The New York Times*, January 30. Available at: www.nytimes.com/2018/01/30/world/africa/cape-town-day-zero.html (Accessed: August 1, 2022).

12 Naishadham, S. (2022) 'How cities in the west have water amid drought,' *Phys.org*, May 24. Available at: https://phys.org/news/2022-05-cities-west-drought.html (Accessed: August 1, 2022).

13 Friedlander, D. (2014) 'Why household size matters' [online]. Available at: https://lifeedited.com/why-household-size-matters/ (Accessed: August 1, 2022).

14 Perry, M. (2016) *New US homes today are 1,000 square feet larger than in 1973 and living space per person has nearly doubled* [online]. Available at: www.aei.org/carpe-diem/new-us-homes-today-are-1000-square-feet-larger-than-in-1973-and-living-space-per-person-has-nearly-doubled/ (Accessed: August 1, 2022).

15 Ratner, M. and Glover, C. (2014) *U.S. energy: overview and key statistics*, June 27 [online]. Available at: https://sgp.fas.org/crs/misc/R40187.pdf (Accessed: August 1, 2022).

16 Litman, T. (2003) 'Economic value of walkability,' *Transportation Research Record*, 1828(1), pp3–11. DOI: 10.3141/1828-01

17 Flint, A. (2014) 'What millennials want and why cities are right to pay them so much attention,' *Bloomberg CityLab*, May 5 [online]. Available at: www.bloomberg.com/news/articles/2014-05-06/what-millennials-want-and-why-cities-are-right-to-pay-them-so-much-attention (Accessed: August 1, 2022).

18 Loh, T.H. and Leinberger, C.B. (2019) 'The economic power of walkability in metro areas,' *Brookings*, July 12 [online]. Available at: www.brookings.edu/blog/the-avenue/2019/07/12/the-economic-power-of-walkability-in-metro-areas/ (Accessed: August 1, 2022).

19 Joha, J. (2019) 'The words and phrases that defined the decade' [online]. Available at: https://mashable.com/article/top-words-phrases-decade-hashtag-culture-war-content (Accessed: August 1, 2022). 'Disruptive Technology' ranked number 10.

20 Christensen, C., Raynor, M. and McDonald, R. (2015) 'What is disruptive innovation? Twenty years after the introduction of the theory, we revisit what it does—and doesn't—explain,' *Harvard Business Review* [online]. Available at: https://hbr.org/2015/12/what-is-disruptive-innovation (Accessed: August 4, 2022). In addition to the article explaining the concept of disruptive innovation, it also discusses how the term has been misunderstood in the media.

21 Katz, B.M. (2015) *Make it new: A history of Silicon Valley design.* Cambridge: MIT Press.

22 Morozov, E. (2013) *To save everything, click here.* New York: Public Affairs.

23 Jaruzelski, B. (2014) 'Why Silicon Valley's success is so hard to replicate,' *Scientific American,* March 14 [online]. Available at: www.scientificamerican.com/article/why-silicon-valleys-success-is-so-hard-to-replicate/ (Accessed: August 1, 2022).

24 Baraniuk, C. (2018) 'Why we shouldn't try to replicate Silicon Valley,' *BBC,* February 8 [online]. Available at: www.bbc.com/worklife/article/20180208-why-we-shouldnt-replicate-silicon-valley-evolution

25 Semuels, A. (2019) 'The world has an e-waste problem,' *Time,* May 23 [online]. Available at: https://time.com/5594380/world-electronic-waste-problem/ (Accessed: August 13, 2022).

26 Waste Electronic and Electrical Equipment Forum (2021) *International e-waste day: 57.4M tonnes expected in 2021* [online]. Available at: https://weee-forum.org/ws_news/international-e-waste-day-2021/ (Accessed: August 13, 2022).

27 Wacks, Y. and Weinstein, A.M. (2021) 'Excessive smartphone use is associated with health problems in adolescents and young adults,' *Frontiers in Psychiatry* [online]. Available at: https://doi.org/10.3389/fpsyt.2021.669042 (Accessed: August 14, 2022).

28 The World Bank (2020) *Life expectancy at birth, total (years)* [online]. Available at: https://data.worldbank.org/indicator/SP.DYN.LE00.IN?most_recent_value_desc=true (Accessed: August 1, 2022).

29 Small, G.W., Lee, J., Kaufman, A., Jalil, J., Siddarth, P., Gaddipati, H., Moody, T.D. and Bookheimer, S.Y. (2020) 'Brain health consequences of digital technology use,' *Dialogues Clin Neurosci,* 22(2), pp179–187. Available at: www.tandfonline.com/doi/full/10.31887/DCNS.2020.22.2/gsmall (Accessed: August 1, 2022).

30 IEA (2019) *Global air conditioner stock, 1990–2050* [online]. Available at: www.iea.org/data-and-statistics/charts/global-air-conditioner-stock-1990-2050 (Accessed: August 9, 2022).

31 Arieff, A. (2016) 'Solving all the wrong problems,' *The New York Times,* July 9. Available at: www.nytimes.com/2016/07/10/opinion/sunday/solving-all-the-wrong-problems.html (Accessed: August 1, 2022).

32 Wright, S. (2022) 'In praise of the humble sidewalk,' *Planning Magazine,* April 27 [online]. Available at: www.planning.org/planning/2022/spring/in-praise-of-the-humble-sidewalk/ (Accessed: August 1, 2022).

33 Surico, J. (2021) 'Downtown Brooklyn's greener, car-free future is taking root,' *Bloomberg CityLab,* May 5 [online]. Available at: www.bloomberg.com/news/articles/2021-10-21/slowly-brooklyn-s-car-free-reinvention-takes-shape (Accessed: August 1, 2022).

34 Mumford, L. (1962) 'The Case Against "Modern Architecture",' *Architectural Record,* 131(4), p157.

35 Papanek, V. (2019) *Design for the real world.* 3rd edn. London: Thames and Hudson, p15.

36 Bourdeau, M. (2022) 'Auguste Comte,' in Zalta, Edward N. (ed.) *The Stanford Encyclopedia of Philosophy* (Spring 2022 edition). Available at: https://plato.stanford.edu/archives/spr2022/entries/comte/ (Accessed: August 1, 2022).

37 Darwin, C. (1859) *On the origin of species.* London: John Murray.

38 Papageorgiou, T. and Michaelides, P.G. (2016) 'Joseph Schumpeter and Thorstein Veblen on technological determinism, individualism and institutions,' *The European Journal of the History of Economic Thought,* 23(1), pp1–30. DOI: 10.1080/09672567.2013.792378

39 Schumpeter, J. (1942) *Capitalism, socialism and democracy.* New York: Harper and Brothers.

40 Nisbet, R. (1980) *The history of the idea of progress.* New York: Routledge.

41 Christensen, C. (1997) *The innovator's dilemma.* Cambridge: Harvard Business Review Press.

42 Morozov, E. (2011) *The net delusion: The dark side of internet freedom.* New York: PublicAffairs.

43 Bijker, W.E., Hughes, T.P. and Pinch, T. (eds.) (1987) *The social construction of technological systems, anniversary edition: New directions in the sociology and history of technology.* Cambridge: MIT Press.

44 Hughes, T. (1987) *American genesis: A century of invention and technological enthusiasm, 1870–1970.* Chicago: University of Chicago Press.

45 Kruppa, M., Schechner, S. and Kendall, B. (2022) 'Google offers concessions to fend off U.S. antitrust lawsuit,' *The Wall Street Journal*, July 8 [online]. Available at: www.wsj.com/articles/google-offers-concessions-to-fend-off-u-s-antitrust-lawsuit-11657296591 (Accessed: July 29, 2022).

46 Slater, C. (1997) 'General Motors and the demise of streetcars,' *Transportation Quarterly*, 51(3) [online]. Available at: https://babel.hathitrust.org/cgi/pt?id=mdp.39015047411684&view=1up&seq=357&skin=2021 (Accessed: July 29, 2022).

47 Ample scholarship has analyzed the mid 20th-century decline of electric streetcar systems across the US, Canada and other nations, some attributing the phenomenon to industrial conspiracy, others to broader cultural and economic forces. Similar arguments have been raised about the demise of General Motor's EV1 electric car in the 1990s in the film documentary *Who Killed the Electric Car?* (2006) Directed by Chris Paine [Film]. Los Angeles: Sony Pictures Classics. Today, as of this writing, automakers are focusing heavily on electric vehicles. In this book, the main interest in electric surface transit is techno-environmental which is most often cited as the best reason to revive these systems in areas where they would most likely succeed. Some cities which once had expansive systems are no longer well-matched to the technology. Los Angeles' Metro, for example, has suffered poor ridership since its introduction despite its popularity with voters. Manvlle, M., Taylor, B. and Blumenberg, E. (2018) *Falling transit ridership: California and Southern California* [online]. Available at: https://scag.ca.gov/sites/main/files/file-attachments/its_scag_transit_ridership.pdf (Accessed: August 1, 2022).

48 Other studies have emphasized the struggle for primacy on the city street and how the automobile prevailed. Norton, P. (2008) *Fighting traffic*. Cambridge: MIT Press.

49 Morris, W. (1890) *News from nowhere*. New York: Vanguard Press.

50 Papanek (2019).

51 Morozov, E. (2014) *To save everything, click here: The folly of technological solutionism*. New York: PublicAffairs.

52 Huesemann, M.H. and Huesemann, J.A. (2011) *Technofix: Why technology won't save us or the environment*. Gabriola Island: New Society Publishers.

53 Vinsel, L. and Russell, A. (2020) *The innovation delusion: How our obsession with the new has disrupted the work that matters most*. New York: Currency.

54 *Wall-E* (2008) Directed by Andrew Stanton [Film]. Emeryville: Pixar Animation Studios.

55 Sax, D. (2018) 'End the innovation obsession: Some of our best ideas are in the rearview mirror,' *The New York Times*, December 7. Available at: www.nytimes.com/2018/12/07/opinion/sunday/end-the-innovation-obsession.html (Accessed: August 1, 2022).

56 Hill, G. (2013) 'Living with less. A lot less,' *The New York Times*, March 9 [online] Available at: www.nytimes.com/2013/03/10/opinion/sunday/living-with-less-a-lot-less.html?pagewanted=2&_r=0&hp (Accessed: December 17, 2022). helped popularize small homes and environmentally conscious living through the blogs Treehugger and LifeEdited.

57 Furnishing Utopia (n.d.) http://furnishing-utopia.com

58 Koolhaas, R. (1978) *Delirious New York*. New York: The Monacelli Press.

59 Ward, B. (2009) *Dr. Space: The life of Wernher von Braun*. Annapolis: Naval Institute Press.

60 Gammon, K. (2021) 'How the billionaire space race could be one giant leap for pollution,' *The Guardian*, July 19 [online]. Available at: www.theguardian.com/science/2021/jul/19/billionaires-space-tourism-environment-emissions (Accessed: August 8, 2022).

61 Singer, C.J. and Raper, R. (1956) *A history of technology, Vol. 2*. London: Oxford University Press.

62 MacNeil, I. (2002) *An encyclopedia of the history of technology*. London: Routledge.

63 Makieła, Z., Stuss, M.M. and Borowiecki, R. (2021) *Sustainability, technology and innovation 4.0*. New York: Routledge.

64 Bieri, A. (2022) *Patents and Professors: The Interdependence between Patent Law, Science, and Research Universities in the United States of America*. Tübingen: Mohr Siebeck.

Chapter 2

The Imperfect Renaissance
of the American City Street

Sidewalks, Bike Lanes and Electric
Surface Transit

Figure 2.1
**Above left: A poster montage of promotional Rural Electrification Administration images from the 1930s.
Home electrification and associated loans for purchasing appliances contributed greatly to the adoption of
electric inventions and designs that have continued to the present day. Source: The Smithsonian Lemelson
Center for Invention and Innovation.**

Figure 2.2
**Above right: Barely a decade later, an earlier energy transition swept North American and other cities—the
eradication of electric surface transit. Pictured in the early 1950s are scrapped Pacific Electric streetcars in
Los Angeles.** *Source: Los Angeles Times.*

It is well-known how the 20th century brought significant changes to the built environment,
notably in the United States where the development of the skyscraper, and later the interstate
highway, transformed cities and their surrounding communities. Whereas walkable cities,
electric public transit and commuter trains had supported the urban fabric through the first
half of the 20th century, the personal automobile eventually prevailed, especially in the US
after the Second World War. Neighborhoods in cities, often historic ones, from St. Louis
to Philadelphia were bulldozed to make space for highway interchanges connecting city
centers and their suburban or exurban outskirts. To be sure, much has been written about
highway infrastructure programs envisioned by Robert Moses in the New York City metropolitan era. Robert Caro's *The Power Broker*, for example, detailed the urban and social impacts
of automobilization.[1] Meanwhile, families, many of them young veterans returning from
military service in the Second World War, moved to new suburban developments at the

DOI: 10.4324/9780367814304-3

perimeter of cities where many were able to afford, often with government loans, single family homes of their own for the first time.[2] Levittown and similar developments sprang up to accommodate rising demand. Automobiles became more affordable for getting around, especially to and from the workplace. As a result, the golden era of the automobile began. Much of the honeymoon phase of the automobile was reflected in the design of cars from the period, especially in the early 1950s and 1960s. Tailfins and other features reflected the optimism of the time; driving, beyond the mundane routine of going to the grocery store, embodied a spirit of wonder, freedom and exploration. Entirely new behaviors and building typologies evolved in parallel; the drive-in movie theater, the drive-in restaurant and then, later, the drive-through restaurant and suburban malls. Millions of acres of parking to support the automotive lifestyle were paved to enable this transformation and new roadside design languages soon followed.[3] While suburban communities had begun to develop prior to the Second World War, enabled by commuter trains and streetcars to reach places of employment (hence the term "streetcar suburbs"), the individualist and liberating quality of the automobile played a pivotal role in drawing people away from cities—to the promise of a better life in suburban Shangri-la. In popular culture, films and literature further romanticized the liberating qualities of driving.[4] Car manufacturers too flooded customers with images of technological bliss through marketing campaigns. General Motors' *Parade of Progress* from 1936–1956 and other exhibits promoted the automobile. Later, the 1964 World's Fair *Futurama* pavilion in New York City presented sleek concept automobiles taking design cues from science fiction and spacecraft.

 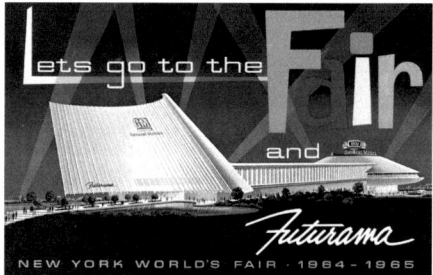

Figure 2.3
Above left: A Pontiac Cirrus exhibited at Futurama during the 1964 New York World's Fair. Concept designs like the Cirrus were meant to stimulate interest and commitment to automotive transportation. Source: GM.

Figure 2.4
Above right: A promotional advertisement for the 1964 New York World's Fair Futurama exhibition. Source: GM.

In time, however, the popularity of the automobile eroded urban centers in their already declining fortunes in the US. Streets were widened to accommodate more traffic and pedestrian sidewalks were narrowed. More highways, bridges and tunnels were proposed and created to support increasing traffic and automotive access. After decades of constructing automotive infrastructure, commuting times increasing, and later, crime rates falling, slowly some American cities began to modestly rebound.[5] Seeking walkability, shorter commutes and public transportation, the desirability of urban life reemerged especially in a few US cities in the 1990s. By the late 2000s, traffic lanes that were paved to support automotive traffic flow were reannexed for pedestrians, bicyclists and bus access.[6]

No longer did the automobile's exceptionalism prevail; aspirational tailfins evolved into utilitarian, boxy minivans and SUVs. What was once considered a democratizing innovation of the 20th century no longer seemed like a human advancement. Tailpipe emissions, lost hours of productivity in gridlock traffic and many other drawbacks, including carbon emissions and air pollution, have soiled the legacy of this modern innovation. Like with many new technologies, their boosters believed their role would be positive and transformative for civilization—a great equalizing force. In other respects, there has been a backlash against the user experience of driving, of being held hostage by automotive-defined environments. Author and geographer Joel Kotkin has described how suburbs, as a consequence, have evolved to reflect many of the advantages of both cities and suburbs, including how people get around for work and leisure. Kotkin's research suggests that these trends will continue.[7] Americans in particular have rediscovered what other countries have held dear and preserved: walkable down towns, public commuter transit and streets where bicycles can be ridden confidently and safely.[8] In New York City, as mentioned before, like in other major American urban areas, streets have been reimagined for pedestrians, bicycles and buses in the 2010s after decades of street widening to prioritize car-traffic. Sequestered bicycle lanes have made transit less dependent on vehicular traffic. Bicycles especially have seen a boom in use which continues today.[9] Streets that were once clogged with vehicles are now populated with outdoor seating and food vendors, taking influence from streetscapes of continental Europe, Asia and elsewhere. To improve quality of life and to address international efforts to meet carbon emission standards, former New York City Mayor Michael Bloomberg made these efforts a centerpiece of his policy agenda. In 2007, Bloomberg even attempted to enact "congestion" pricing, inspired by similar efforts in London, in a further effort to roll back automobile traffic in the city center.

Many of Mr. Bloomberg's efforts were the culmination of decades of work to resuscitate the city from financial duress and the effects of the automobile. In the 1960s, part of the SOHO neighborhood in lower Manhattan was nearly bulldozed for a highway envisioned by Robert Moses, who led the Triborough Bridge and Tunnel Authority. At the time, SOHO was blighted but full of historically significant 19th-century manufacturing buildings. Moses' Lower Manhattan Expressway, supported by powerful interests, was presented as an attempt to modernize the city, which ultimately failed.[10]

This watershed moment of urban preservation, influenced by the writings and activism of urban planner Jane Jacobs,[11] began a decades-long reversal to a quasi-modern pedestrian urban experience that looked more like 1921 than 1961.[12] The outcome of America's urban renaissance has also helped diminish energy use and air pollutants as people choose public transit, bicycles or their own legs to get around. In parallel to these urban developments, bicycles and modes of public transportation have reemerged as well, some with greater success than others.

While bicycles have remained in use around the world since their invention in various countries in the 19th century, bike sharing platforms and electric assist hybrids offer new incentives to adopt and adapt this older technology for traditional and new uses. American bicycle ridership has surged especially during the Covid-19 pandemic.[13] In many ways, bicycles are a quintessential example of reimagining and redesigning an older technology to make it relevant in the present for current users and contexts. Despite earlier variants, notably the 1819 velocipede, the first commercial chain drive pedal powered bicycle, introduced around 1885 by Rover in the UK, underwent few major overall changes through today. Materials, pneumatic tires, seats and other details have evolved, but the basic configuration has remained. The design of bicycles has, of course, diversified into numerous categories in the past 135 years: E-bikes, racing bikes, mountain bikes, folding bikes, cruisers and more. Like the evolution of the chain-driven two-wheeled original, additional alternative technologies could also benefit from similar design attention to be successfully reintroduced for users in today's context.

Along with the resurgent popularity of bicycles, the dedicated bicycle path has, as mentioned, also experienced a renaissance. Beginning in the 1890s, dedicated bicycle paths and even elevated structures were built to support the bicycling craze which lasted into the first decade of the 20th century. Dedicated bicycle paths, some with tolls, were constructed in Los Angeles, Pasadena, New York, New Jersey and other locations. These paths were created to improve the riding experience, mainly from muddy dirt roads and obstructions like pedestrians, streetcars and the occasional automobile. Over time, most of these roadways were abandoned. Those that ran alongside roads were absorbed into streets when they were paved and overhead bicycle paths were dismantled. As the auto-mobile began to dominate roads, bicycle usage declined in the West, especially in the US, although around the world they remained in wide usage and continue to be today. Other countries which have prospered economically have begun to substitute cars for bicycles, notably China which continues to be the most populous country on earth.

Recent History of Protected Bicycle Lanes

In the late 1960s and 1970s, protected bicycle lanes began to be reintroduced especially in Europe as oil prices and traffic deaths spiked. Beginning in Sweden and Finland, bike lanes began to be reconstructed in the Netherlands and Denmark as well. Gradually, transportation planners began to build protected lanes, often in college towns with large populations of cyclists. By the mid 2000s, even large congested cities like New York that once had early bike lanes eventually followed suit. Like other cities around the world, New York also launched a bike sharing platform during that time to further encourage the use of this new infrastructure.

Figure 2.5
Above left: A 1900 photo of the elevated California Cycleway in Pasadena, California. Riders could cycle without obstacles for a toll. Source: Putnam; (Firm.) Southern Pacific Railroad Company.

Figure 2.6
Above right: A typical early 20th-century protected bike lane in an unidentified location in western Europe. Source: Markus Spiske.

Reintroducing Urban Electric Streetcars with Mixed Success

Electric-powered public transit has also undergone a relatively recent renaissance in the US just as bicycling and pedestrianism. The electric streetcar specifically is perhaps one of the best examples of a beneficial alternative technology that has been usefully reintroduced after a long hiatus. This is especially true in the US, which once had dozens of streetcar systems, few of which survive. Streetcars, also known as "trolleys" or "trams," were first developed as an enhancement to horse drawn precedents. One of the earliest commercially successful electric systems was demonstrated by Werner von Siemens in 1879. Driven by a relatively small 2.2 kW series-wound motor, the early streetcar reached a speed of 13 km/h. In a period of four months, the train carried 90,000 passengers on a 300-metre-long circular track. Electricity was supplied through a third insulated rail between the tracks. A contact roller was used to connect to the electric energy source below on the street. Just afterward, in 1881, an electric streetcar line opened for passenger service in Lichterfelde, on the outskirts of Berlin.[14]

Without automotive traffic to compete with, the popularity of streetcar systems quickly spread elsewhere, as city officials compared notes about the methods and means of their construction and maintenance. In the US, a few cities competed to introduce electric streetcars. Richmond, Virginia introduced one of the first continuously operating electric streetcar systems. Opening for service in 1888, it was built on the technological developments of inventor Frank Sprague, who had previously worked on electrical technology in Thomas Edison's research lab in New Jersey. By the end of the 19th century, major cities around the world had developed electric-powered streetcar systems; they built new or replaced horse drawn, cable or steam-powered alternatives. Usually they were powered by overhead wires, but a few used a third insulated rail like Siemens' original prototype. At one time dozens of streetcar systems in all states were constructed in population centers as large as New York City and as small as Johnstown, PA (population ~63,000 in 1950). Across the US, from San Diego to Boston and most every urban center in between, residents could travel between downtowns and to outlying areas for a modest fare, in an energy-efficient electric vehicle. Some cities, like Roanoke, Virginia even constructed local hydro-electric plants to produce electricity for streetcars.

Only a few decades later, by the 1920s, American industrialist Henry Ford had improved manufacturing assembly lines enabling the personal automobile to be more affordable. Eventually, this development contributed to the automobile's primacy in American mobility for decades to come. By 1948, more than half of the electric street-car systems had been abandoned in the US, replaced most often with diesel-powered buses. A similar fate unfolded in the UK and some other countries. In El Paso, Texas the International Tramway, used to cross the Mexican border to Ciudad Juarez, remained in service until the 1970s.

Streetcars were generally popular with riders, but disliked by motorists for blocking traffic. Buses could at least pull out of traffic when picking up or dropping off passengers. Some well-publicized conspiracy theories have argued that an industry consortium with interests in the sale of petroleum-powered buses plotted to systematically acquire and dismantle American and Canadian streetcar lines, so that they could be replaced with the petroleum-powered buses that they collectively manufactured and sold. Others have argued that fixed fares and unfair infrastructure obligations on streetcar companies to repair streets around their rails made it impossible for them to compete with private bus lines.[15] As a result of the decline of electric surface transit systems in favor of buses, tailpipe emissions increased in the cities where they had been used.

By the 1970s, evolved versions of streetcar systems began to reemerge in American cities, not long after their predecessors had been dismantled. Grid lock traffic, rising gasoline prices and steadily increasing commuting times were all contributing factors. Commuter train ridership grew in turn. Although new generation streetcars differed from old systems in a few significant ways, most new systems referred to as "light rail" were constructed with dedicated right of ways so they could go faster by not competing for street space with automotive traffic. In some cases, these systems would leave the city center to destinations in the suburbs. Many were built on portions of rail right of ways from abandoned early streetcar systems. Some streetcar systems were subsequently restored or replaced by underground rail systems. Washington D.C., for example, had an elaborate network of streetcars, which were closed by 1962, then replaced in part by the DC Metro in the late 1970s. Today the Metro network continues to expand. In contrast, streetcar systems that emulated original urban in-street systems in cities that had grown and sprawled performed less successfully—for example, Washington D.C.'s DC streetcar, begun in 2006, competes with cars and buses but also serves no foreseeable purpose; it has riders probably because it is free. Nevertheless, new light rail streetcar systems across the nation improved inner city air quality, quality of life for residents and commuters while reducing traffic loads and energy consumption. Portland, Oregon's MAX transit system, introduced in 1986, used former streetcar and regional interurban right of ways that had closed in 1950 and 1958 respectively.[16] Unlike earlier systems that competed with traffic, much of MAX's right of way is private, especially outside of downtown, unencumbered by automobile traffic. In Europe, Mexico and elsewhere, many streetcar systems were also abandoned. In the UK, a country which once had numerous streetcar and trolleybus systems, few have survived, but some have been rebuilt. Construction of underground metros have expanded as well. On the other hand, systems in parts of Asia and especially in continental Europe have remained intact, succeeding in part from stronger federal financial commitment.

Newer streetcar systems have had mixed success despite their many benefits to riders, energy efficiency and local air quality. Much of their failure can be attributed to poor planning. Some have failed to attract enduring ridership despite their popular support in ballot measures which raised taxes to finance them.[17] Newer systems in Los Angeles, Philadelphia and Washington, D.C. all suffer from lackluster revenue. Los Angeles, for example, had one of the largest privately owned electric streetcar and interurban systems (Pacific Electric), lasting for decades before closing completely in the 1960s. The newer system, whose construction began in the late 1990s, has not been as widely used as originally hoped.[18] Their limited success can be partially explained by the changing urbanism of the city itself. Since the Second World War, Los Angeles and connected counties have become famously decentralized, such that commuter populations no longer travel from the perimeter of the city to a central urban downtown. As such, L.A.'s limited number of Metro lines are challenged to bring most riders close to their desired destination.

In other cities, streetcar lines have been reintroduced or built new, some to attract tourists and promote economic development. Philadelphia's "vintage" trolley, route 15, served west Philadelphia to the eastern Philadelphia neighborhood of Northern Liberties using remanufactured PCC streetcars built in the 1940s. Competing with traffic, narrow streets and double parked cars in a dense urban area, the line struggled and the old rolling stock needed extra maintenance and care. Despite solid ridership, the line was temporarily suspended in 2020, citing maintenance problems. Many have argued that buses would better match the needs of local ridership, including allowing the line to extend to an intermodal transportation terminal on 69th street in west Philadelphia. Still, the electric wires in place could be used by modern trolley buses to minimize exhaust emissions.[19] The basic learning here is significant and reflects other examples presented in this book: reintroducing alternative technologies must be analyzed and reinterpreted to meet *present* needs, user preferences and social contexts and, in this case, enhance transportation options rather than competing with existing ones, like the automobile.

Figure 2.7
Above left: A Portland streetcar running on the city's new system installed in 2000–2001. Here the streetcar competes with automotive traffic as opposed to running on a dedicated right of way. Streetcars that run on their own right of way are more costly but generally faster. Examples of dedicated rights of way can be found in Boston, San Francisco and elsewhere. Source: Steve Morgan (CC BY 3.0).

Figure 2.8
Above right: A San Francisco streetcar running in a section of private right of way and tunnel, 1967. This line is still operating today with a large underground section downtown. Source: Marty Bernard.

Notes

1 Caro, R. (1974) *The power broker*. New York: Knopf.
2 Altschuler, G. and Blumin, S. (2009) *The GI bill: The new deal for veterans*. Oxford: Oxford University Press.
3 Venturi, R., Scott Brown, D. and Izenour, S. (1972) *Learning from Las Vegas*. Cambridge: MIT Press.
4 Kerouac, J. (1957) *One the road*. New York: Viking Press.
5 Glaeser, E.L. and Shapiro, J. (2001) 'Is there a new urbanism? The growth of U.S. cities in the 1990s,' *National Bureau of Economic Research* [online]. Available at: DOI 10.3386/w8357
6 Friss, E. (2019) *On bicycles: A 200-year history of cycling in New York City*. New York: Columbia University Press.
7 Kotkin, J. (2017) *The human city: Urbanism for the rest of us*. Evanston: Agate Publishing.
8 Couture, V. and Handbury, J. (2019) 'Urban revival in America, 2000 to 2010,' *National Bureau of Economic Research* [online]. Available at: www.nber.org/papers/w24084 (Accessed: August 5, 2022).
9 Pucher, J. and Buehler, R. (eds.) (2012) *City cycling*. Cambridge: MIT Press.
10 Jackson, K.T. (2007) 'Robert Moses and the rise of New York: The power broker in perspective,' in Ballon, H. and Jackson, K.T. (eds.) *Robert Moses and the modern city: The transformation of New York*. New York: W.W. Norton & Company, pp. 67–71.

11 Jacobs, J. (1961) *The death and life of great American cities*. New York: Random House.

12 Sax, D. (2018) 'End the innovation obsession: Some of our best ideas are in the rearview mirror,' *The New York Times*, December 7. Available at: www.nytimes.com/2018/12/07/opinion/sunday/end-the-innovation-obsession.html (Accessed: August 1, 2022).

13 Bernhard, A. (2020) 'The great bicycle boom of 2020,' *BBC* [online]. Available at: www.bbc.com/future/bespoke/made-on-earth/the-great-bicycle-boom-of-2020.html (Accessed: August 5, 2022).

14 Electric streetcar technology appears to have emerged in Europe in the late 1870s and 1880s.

15 Mallach, S. (1979) 'The origins of the decline of urban mass transportation in the United States, 1890–1930,' *Urbanism Past & Present*, 8, pp1–17. Available at: www.jstor.org/stable/44368292 (Accessed: August 1, 2022).

16 Thompson, R. (2012) 'Portland street car system,' *Oregon Encyclopedia* [online]. Available at: www.oregonencyclopedia.org/articles/portland_streetcar_system/#.YDfHIS2cbao (Accessed: August 1, 2022).

17 Metro. (2016) *Measure M Voters passed Metro's no sunset transportation ballot measure with 71.15% support* [online]. Available at: www.metro.net/about/measure-m/ (Accessed: August 5, 2022).

18 Manville, M., Taylor, B. and Blumenberg, E. (2018) 'Falling transit ridership: California and Southern California,' *UCLA Institute of Transportation Studies* [online]. Available at: https://scag.ca.gov/sites/main/files/file-attachments/its_scag_transit_ridership.pdf (Accessed: August 5, 2022).

19 Saska, J. (2018) 'Overhauling its bus network may be on SEPTA's schedule soon,' *WHYY*, February 18 [online]. Available at: https://whyy.org/segments/overhauling-its-bus-network-may-be-on-septas-schedule-soon/ (Accessed: August 1, 2022).

Chapter 3

Heating and Cooling

Alternatives for Thermal Comfort

On the hottest, most humid days of summer one might pause to consider how anyone survived before the invention of air conditioning. Air conditioning (AC) offers human comfort in warm regions of the world that can afford it, or at least where the electrical grid is robust enough to provide it. By compressing air to make it cooler and drier, its use has become an expected amenity in the wealthiest precincts of human civilization. Some have argued that it is now a necessity to live in some regions of the world. As the populations and economies of China, India and other warm climates have grown, demand for AC has expanded, and will continue to grow, creating the need for dramatically more energy to meet demand.[1] Similarly, as the internet has flourished in the past decades, air-conditioned web server farms have proliferated as well.

It was only in 1902 that the first practical air conditioner was developed commercially. Conceived by American Willis Carrier, AC was initially used to reduce humidity in industrial printing facilities to prevent paper from warping and sticking together. Only shortly afterward, the technology was adapted to cool spaces for thermal comfort. At first, air conditioning became widespread in the US during the mid 20th century. Since then, the popularity of air conditioning has spread around the world—to Europe, South America, the Middle East and Asia. In 2022, over 90% of Japanese and US households have air conditioning, compared with 16% in Mexico and 5% in India.[2] Unsurprisingly, AC also accounts for a staggering amount of energy use: between 12–17% of electricity use in the American home, averaged over the year even though so many regions use it for only a few months during warmer seasons.[3,4] Production of the electricity to provide air conditioning also has an environmental cost in greenhouse gas emissions. In the US, only a small fraction of electricity generation comes from renewable sources like wind, hydro power and solar power, the rest coming from natural gas, coal and nuclear plants.[5]

Without a doubt, air conditioning provides welcome thermal comfort, but it is often used indiscriminately. In so many commercial buildings now, inhabitants often can't control how much cooled air enters a room—sometimes workers keep sweaters in their office when AC is used over zealously. This has led architects and building scientists to consider alternative ways to mitigate or balance the use of AC with other technologies, including passive breeze features, shading and insulation. Searching for ways to improve the efficiency of AC has also become a top priority by the International Energy Agency.[6] This chapter will consider alternatives used to provide thermal comfort, recognizing that the almighty AC is an entrenched technology. Alternatives must be assessed carefully within the cultural context of the present day, to be accepted and ultimately adopted. As passive cooling expert Dr. Susan Roaf suggests, notably

DOI: 10.4324/9780367814304-4

in reference to the cooling abilities of ancient, passive Persian Wind Catchers, "[we must] not follow the stereotypical thinking of 'conventional wisdom' but to rethink each design decision clearly for ourselves."[7]

Especially in architecture, a great deal has been written about air conditioning and its influence on the design of the built environment.[8] Moreover, some have even argued that some types of modern architecture, specifically the aesthetics of the glassy International Style curtain wall, would never have flourished without the invention of air conditioning.[9,10] The greatest early exemplars of this style can be seen in the influential designs of Mies van der Rohe, Gordon Bunschaft, Natalie de Blois, Philip Johnson and many others. Mies van der Rohe's crystalline Farnsworth house in the Chicago suburbs represents a *ne plus ultra* of this design aesthetic, although the house was not outfitted with AC in the beginning. An elegant, glass prism with light steel framing, it is broadly considered one of the most significant milestones of 20th-century architectural design. At the same time, Dr. Edith Farnsworth, who commissioned the experimental house and used it as a vacation retreat until 1972, was deeply disappointed in the outcome of the design even though she approved the plans and oversaw its construction. In 1951, when completed, Mies and Farnsworth engaged in a bitter legal battle over perceived failures of the design and for non-payment of fees. Among Farnsworth's grievances, the very sealed glass interior space which garnered so much critical praise also heated up tremendously from solar heat gain during the hot humid Illinois summers (air conditioning was installed later in the mid 1970s).[11] While this glass enclosure was well suited to the milder, drier southern Californian environment where the famous glassy Case Study Houses were

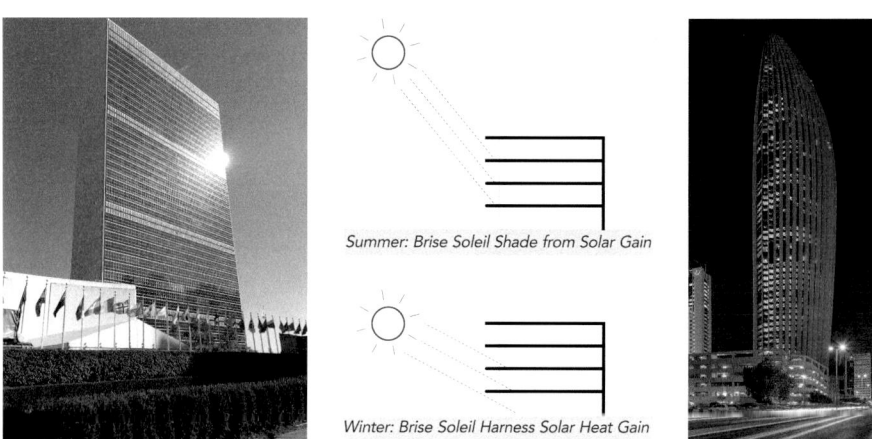

Summer: Brise Soleil Shade from Solar Gain

Winter: Brise Soleil Harness Solar Heat Gain

Figure 3.1
Above left: The 1950 International Style United Nations Secretariat, New York City. A design consortium led by Le Corbusier with Wallace Harrison and Oscar Niemeyer. Today numerous buildings designed like this iconic original are cooled using air conditioning. Le Corbusier favored a scheme with Brise-Soleil to mitigate heat gain. Source: Veni Markovski (CC BY-SA 4.0).

Figure 3.2
Above center: A diagram showing how Brise Soleil manages heat gain on glass façades during warmer summer months. Source: Brook Kennedy.

Figure 3.3
Above right: In 2022, architecture firm Foster + Partners revealed the National Bank of Kuwait skyscraper that celebrates vertical brise soleil along the entire length of the façade. Source: Nigel Young/Foster + Partners.

constructed,[12] the Farnsworth house certainly proved that there was nothing internationally appropriate about this architectural style at all. Later, International Style skyscrapers that lined important corporate boulevards in New York, Chicago and other major cities found that passive design strategies like shading and sun reflecting glass were essential to cut down on cooling costs associated with pure glass curtain walls. The United Nations Secretariat Building in New York City, designed by a consortium of architects led by Le Corbusier, has come to epitomize the shortcomings of the International Style, especially given the lofty aspirations of the UN's Sustainable Development Goals cited in the previous chapters.

Le Corbusier originally proposed façade shading features called Brise Soleil ("sun breaker" in French) to minimize solar gain on the glass windows just as he had incorporated in buildings in warmer climates in France and India. Unfortunately, this feature was eliminated in the UN building's design. Le Corbusier bemoaned, "my strong belief is that it is senseless to build in New York City, where the climate is terrible in summer, large areas of glass that aren't equipped with brise-soleils," he said. "I say this is dangerous, very seriously dangerous."[13] The results of the decision to forego the Brise Soleil had financial and environmental consequences. In spite of the building's blue-green glassy beauty, the UN Secretariat has a substantial air conditioning bill, in part because of the roughly east and west facing all-glass façades. According to recent figures, the UN building, a monument to global civilization and sustainable futures, now suffers from a yearly cooling bill approaching $10 Million.[14]

Smart Home Technologies

"Policy makers think you can solve energy and building problems with gadgets [but] You can't. As global temperatures continue to rise, we are going to continue to squander more and more energy on keeping our buildings mechanically cool until we have run out of capacity."[15]

—Dr. C. Alan Short, Cambridge University

More recently, highly promoted smart home technologies, like the Google Nest thermostat, were developed to help home owners and businesses manage indoor temperatures, reduce their energy bills and for those concerned, reduce their environmental footprint. By anticipating when residents are home, those types of thermostats can save some energy by turning the AC (or heat) off when no one is home. Artificial Intelligence (AI), connected to smartphone-based location tracking, can help anticipate when home owners are returning home and can start cooling ahead of their arrival. They can also be activated manually if the home owner remembers to do so. Yet despite the cleverness of these AI gadgets, remote controls and other benefits, AC combined with inefficient building practices—including poor insulation, ample solar gain or lack of shade, passive and mechanical ventilation and many others design features—can result in significant collective energy waste in the home. In other words, Smart thermostats can't adequately address *the root cause* of energy-inefficient building design. So how did human beings keep cool before air conditioning? Did they simply suffer silently? Overall, humans constructed buildings very differently in the absence of AC, depending on local environmental factors including temperature. The earliest known interventions date from thousands of years ago. But the lessons and abilities of these alternative technologies are more relevant today than ever and have been of increasing interest to practicing architects in the 21st century.

Traditional Building Technologies: Windows, Awnings, Porches and More

Before the arrival of air conditioning in the early 20th century, buildings used a variety of passive and low or no energy methods to help preserve thermal comfort indoors. Of course, thermal comfort, or what is considered a "comfortable" temperature, is subjective. Put in more concrete terms, passive techniques can reduce indoor temperatures several degrees Fahrenheit compared with the outdoors. On the other hand, with access to unlimited electricity, air conditioning can reduce indoor temperatures far more such that a humid, 100-degree Fahrenheit day can feel cool and dry. Depending on several factors including the relative humidity of the local climate, different passive techniques used to be more commonplace in architecture to improve comfort and have been a frequent subject of discussion in passive and net-zero architecture and building trades, notably through standards like Passive House, LEED and others.[16] Many of these energy saving techniques (like using effective insulation) are especially commonplace in Europe where energy costs are higher and effective traditional practices have endured.

The Noble Window Shutter

The window shutter plays a unique role in architecture around the world. Earlier, in the preface, this book discussed how shutters remain actively used to prevent solar heat gain indoors during the summer months. On the other hand, there are numerous other ways to manage solar gain indoors with shades, curtains, venetian blinds and even indoor shutters. While these help, they don't work as well because they still allow light to first pass through the glass window, causing solar heat gain, as experienced in a greenhouse. External shutters, by contrast, keep light from entering the building in the first place. In the US today, window shutters have been demoted to an inoperable decorative accessory, sometimes for windows too wide for the shutters to even cover. Sometimes they are preserved on historical buildings like Thomas Jefferson's Academical Village on the University of Virginia Campus, now a UNESCO World Heritage Site. Here, along the shaded walkways, shutters are preserved for historical accuracy, but they can also be used to help reduce temperatures indoors. As insect-repelling window screens have become more popular, window shutters became harder to close from inside. Surely window manufacturers and architectural product designers could find more ways to either make shutters viable, through easier indoor operability, improved usability and refined contemporary aesthetics. On the other hand, in many countries around the world, shutters are, of course, still widely used along with newer automated variants to achieve the same or improved temperature control.

In other cases, shutters are avoided for their old-fashioned connotations, often in modernist architecture. Some contemporary and modernist buildings employ newer shutter and shade-inspired design features to provide the same functional role and passive cooling assistance as traditional shutters did in premodern buildings. In stark contrast, the uniform glass façades of mid-century International Style skyscrapers, like the previously mentioned UN Building, avoided these features, relying instead on AC, UV and heat-resistant glass and internal shades to mitigate sun load, heat gain or a "greenhouse" effect.[17]

Figure 3.4
Above left: Shutters old and new in use in Aarau, Switzerland. The older shutters on the left are functional, maintained and practical. Those on the right are more contemporary louvered shutters. Source: Brook Kennedy.

Figure 3.5
Above right: An interior strap for raising or lowering external louvered shutters from the inside without opening the windows or screens. Source: Brook Kennedy.

Figure 3.6
Above left: A glass curtain-walled office building in Aarau, Switzerland has two envelopes: in between the outer glass façade and the building's insulated windows is an automated blind system that lowers using sensors under heavy sun load to prevent heat gain indoors. Although for some it might not be as visually striking as the UN Secretariat or Lever House glass box aesthetic, the building has no air conditioning. Source: Brook Kennedy.

Figure 3.7
Above right: Elcano housing block in Madrid, Spain by architecture firm FRPO transforms a façade into an elegant, poetic grid of traditional shutters. Source: ImagenSubliminal (Miguel de Guzman + Rocio Romero).

As an alternative to shutters, many other techniques are available. Architect Daniel Barber's book, *Modern Architecture and Climate: Design Before Air Conditioning*, documents how renowned architects from Le Corbusier, Richard Neutra, MMM Roberto, Lúcio Costa, the Olgay brothers, among others, employed many kinds of passive techniques to respond to local climates in resourceful ways before the arrival and predominance of central air conditioning.[18]

Awnings, Air Flow and Evolving Architectural Tastes

Beyond shutters, other common passive features, like awnings, can help keep spaces cooler. By providing shade, they can reduce solar gain but without blocking out light completely like many shutter designs. Consisting of simple framed structures, with a fabric skin, awnings come in fixed and deployable variants that can be used as needed. Later static designs, made of aluminum, appeared in the mid-century. The concept of deployable awnings dates to at least the Roman Empire where they were used for open air colosseums and other purposes. Today, in central and southern Europe, modern designs for building awnings are popular and widespread, specified on buildings of all kinds and in divergent styles, from traditional to contemporary. Automated, technical versions have also emerged, some that fold out automatically when heat gain is detected and even some have solar photovoltaic panels to gather electricity. In contrast, they are seldom specified in American buildings today despite their practical, energy saving simplicity. While the abundance and low cost of air conditioning is one explanation, like the shutter, they are often dismissed as old-fashioned historical period details. Still, awnings could be redesigned for contemporary acceptance and used more extensively again to help keep buildings cool, while lowering air-conditioning costs.

Figure 3.8
Above left: A US patent image of a retractable awning found on American shop fronts were commonplace on 19th- and early 20th-century US commercial and residential buildings. Later they disappeared as air conditioning proliferated and architectural tastes evolved. Source: USPTO.

Figure 3.9
Above right: Contemporary buildings in continental Europe often use awnings like these in Denmark. Source: Victoire Joncheray.

Harnessing and Creating a Breeze: Wind Towers and Wind Catchers

Wind towers, developed in Persia and China as much as 3,000 years ago, are chimney-shaped towers used for capturing wind flows for cooling purposes. Overall, they can provide buildings with moderate temperature control.[19] When used in conjunction with underground irrigation canals (Qanats), desert buildings in northern Iran and the Middle East are able to lower ambient temperature indoors by as much as 8–10 degrees Fahrenheit during the summer.[20] By capturing and channeling wind flows above the roofline of most buildings, air is circulated through the interior which in turn expels hotter air outside through a companion exhaust duct.

Figure 3.10
A wind catcher in Iran used for passive cooling. The towers capture passing wind introducing cooler air indoors. Sometimes subterranean irrigation canals, known as Qanat, helped produce an evaporative cooling effect capable of reducing the temperature indoors nearly 10 degrees Fahrenheit. Source: Julia Maudlin.

Despite their age, wind catchers continue to be used successfully in the Middle East. Today and in the future, wind catchers could be designed and integrated into new buildings to provide passive cooling for residents, especially in hot, arid regions around the world. Beyond Persia, the United Arab Emirates and Oman, the western United States could host this ancient technology for energy saving benefits. Several examples already exist in Colorado, California and Utah, one at a park service visitor's center building in Zion National Park,[21] and a second was incorporated into a house designed by Design Build BLUFF and the University of Colorado at Denver.[22] This house also features Native American inspired rammed-earth construction which further insulates cooler air indoors from hotter temperatures outside. At the time of the writing of this book, these buildings' energy performance has not been studied or measured to the level of the ancient Persian precedents, but this is clearly an important next step. If impactful, given the overall cost of construction, why couldn't wind catchers be more widely incorporated into commercial buildings and homes? Perhaps even a new architectural typology could emerge—one that moves away from the nostalgic, climate mismatched architectural styles often found in new housing developments. In Kenya, the 2022 Pritzker Prize-winning architect Diébédo Francis Kéré designed a passively cooled series of buildings belonging to an Information Technology training center called the Startup Lions Campus. Using similar design and technology techniques as the Persian wind catchers (and bioinspiration from local termite mounds) the buildings are able to passively maintain cooler temperatures for the educational activities indoors.

 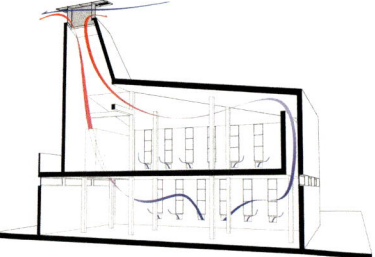

Figure 3.11
Above left: Startup Lions Campus, 2021, in Kenya, by Kéré Architecture employs cooling chimneys based on termite colony mounds and simultaneously the wind catchers of the Middle East. The distinctively shaped chimneys provide a salient visual feature to identify their energy saving ability. Source: Kéré Architecture.

Figure 3.12
Above right: A section view of the Startup Lions Campus showing the passive movement of air indoors. Source: Kéré Architecture.

Dog Trot Houses—Designing Breezeways

Capturing wind flow to promote passive cooling can be achieved in additional ways. Today, in the American southeast, a climate legendary for sweltering summer heat and humidity, regional inhabitants had to be especially resourceful before the advent of electrically powered mechanical cooling. Once air conditioning arrived, more people began to move south. In fact, recent population growth in the southern latitudes of the US have even been tied to the rise and proliferation of air conditioning.[23] Before homes could be cooled at the flip of a switch, houses in the southeast evolved to keep cool. Dog Trot houses, for instance, offer some insight indicating how this was achieved. While their origin is uncertain, a common theory is that they originated from haphazardly connecting two log cabins under one roof sometime in the 18th or 19th century. In between the two dwelling halves, a narrow breezeway would funnel airflow through the hollow space all day long. Sometimes, inhabitants would congregate or eat in this open central space. Along with porches, higher ceilings, shutters, awnings and other techniques, Dog Trot houses were a popular form of vernacular building design used across the socio-economic spectrum. However, few original structures have survived. The 1849 Autrey House in Dubach, Louisiana is a fine, well-preserved example for inspiration.

Today, architects have built contemporary structures drawing from the designs of Dog Trot houses like these.[24] In Europe and Africa, Jean Prouvé's La Maison Tropicale used many overlapping ventilation techniques found in Dog Trot houses although there is no formal lineage between the two. While there is no dedicated central breezeway in La Maison Tropicale, the sliding opening walls allow air flow to pass through the house.

Figure 3.13
Above left: The 1818 Looney House, a Dog Trot example in Ashville, Alabama, US. Note the characteristic breezeway through the center of the house to allow air flow during summer months. Source: Chris (CC By 2.0).

Figure 3.14
Above right: A 20th-century replica of Jean Prouvé's La Maison Tropicale in Brazil. Although not related to the Dog Trot houses of the American south, the design uses similar techniques to promote breezes and air flow for cooling thermal comfort. Source: Gabriel Fernandes.

Fresh Air and the Enduring Genius of the Operable Window

Opening glass windows, a technological staple of architecture, is easily taken for granted and has a more nuanced history than simply allowing natural light and air indoors. Certainly, windows can keep an indoor climate partially insulated from an uncomfortably hot or cold outdoor environment, but, as partially discussed already, they can also lead to solar heat gain, essentially creating a greenhouse effect indoors. Newer double- and triple-glazed windows have helped improve the insulating qualities of windows to prevent these warming side effects.

Some evidence suggests that glass and other materials were used to cover outer wall openings in Roman Egypt as early as 100AD. Later, in the middle-ages, stained glass was introduced in ecclesiastical buildings, notably the early Gothic cathedrals in the 12th century to create an emotive effect. Only in the 17th century did optically clear glass windows appear more commonly in residential houses. Johannes Vermeer's famous paintings, including "The Milkmaid" (1657–1658), highlight the novelty of this new architectural feature and the quality of light it brings into an interior space.

However, the English word "window" was likely derived as much from its use to allow air passage inside and out. The possible root of the word comes from the old Norse "vindauga" literally meaning "wind eye"[25] or presumably a means of allowing air through while being able to look out as well. Similarly, the Latin root "Fenestra" influenced the word for window in continental languages (*Fenster:* German, *Fenêtre:* French and *Fönster:* Swedish). In Biological terminology it is used to describe a pore—a feature that allows material to pass through, including air. Yet even though the window's role as a passage for fresh air predates that of the now ubiquitous glass covered window, its role as a means of admitting fresh air and expelling stale air is now secondary in many countries including the US.[26] As Heating, Ventilation and Air Conditioning (HVAC) systems have become architecturally predominant, the vestigial utility of an openable window has seemingly been demoted. Concerns about burglary, safety and other issues have only compounded their demise in contemporary architecture. Even buildings that have retained operable windows, locks, restraints and even paint have limited the once important role of the opening window. As an exception, in Germany and other adjacent continental European countries, the practice of *Luften*, essentially simply opening windows and introducing fresh air, dates back centuries. Observed as a daily ritual, *Luften* is generally carried out twice daily in the morning and evening, often after cooking to rid interior spaces of smoke, cooking odors and foul air. During the height of the Covid-19 pandemic, the practice of *Luften* was passionately debated in Alemannic countries—pitting public health officials and traditionalists against environmentalists who point out the energy waste associated with opening an otherwise well-insulated home to the elements.[27] It's a complicated issue! The Russian *Fortochka*, however, proposes a compromise—these window designs found in Russia and Ukraine have a small openable window inset into a larger one that could be cracked open during the coldest winter months without as much heat escaping. Building scientists have also proposed further efficiency enhancements with louvers.[28]

Nevertheless, it is well established that access to fresh air is an essential condition for human health and has been for thousands of years.[29,30] In the 19th century, as countries in the northern hemisphere industrialized, where skies and cities were filled with smoke, particles and other by-products of industrial production, public concern about the impact of air quality on human health increased. The emergent "clean air movement," comprised of designers, architects and landscape architects, began to imagine ways to improve the quality of life in cities that were largely affected by poor air quality. In the 1850s, Frederick Law Olmstead, the leading park designer of his generation, envisioned Central Park in Manhattan to serve as the "Lungs of the City."[31] Later, New York City passed early housing regulations embodied in the New Law tenement apartment buildings that required air shafts and windows to admit fresh air inside working-class, mostly immigrant apartments.[32] Ebenezer Howard's *Garden Cities of To-Morrow* (1898) championed less dense urban and suburban developments which prioritized tree planting and open space in the name of health and quality of life.[33] Altogether, design efforts from this period sought to prioritize improving air quality, especially indoors. Then just as now, designers understood that fresh air, ventilation, windows, trees and other interventions were able to help reverse the effects of poor air quality indoors. Consequently, exposure to fresh air has become synonymous with healthy activity. "I'm going to get out and get some fresh air" became English shorthand with taking a healthy break. Even so, outdoor air quality remains a challenge in cities around the world, resulting from industry and tailpipe emissions. In many urban areas, from Beijing to New York, health experts recommend residents remain indoors when air quality is particularly poor. Electric vehicles have been promoted in part for these reasons as well.

Fresh Air, Disease Transmission and Convalescence

In the late 18th and early 19th centuries, health officials became convinced that fresh air could be an effective tool for preventing the transmission of some communicable diseases.[34] By the late 19th century, building design evolved to leverage natural ventilation indoors to achieve this aim. In *The Recovery of Natural Environments in Architecture: Air, Comfort and Climate*, researcher Alan Short discusses advanced methods of ventilating buildings through the 19th and early 20th centuries, including the British Houses of Parliament, and the design of several hospitals.[35] Of special focus is the design of the first Johns Hopkins Hospital in Baltimore, Maryland, US (1873–1889), by John Shaw Billings. Here, the hospital relied on central air shafts to minimize disease transmission without modern HVAC systems, filtration and energy. Other examples of advanced ventilation design, like the Dresselhuys Pavillion of the Zonestraal Sanatorium in Hilverstrum, Netherlands (1950–1974), by Jan Duiker, incorporated open-air porches in the recovery wing for use in more temperate months of the year.[36] Many of these features also helped promote indoor cooling in warmer months.

Designing to promote natural ventilation expanded into other types of institutional buildings. At the onset of the 1918 Influenza pandemic, public health officials recommended simply opening windows in buildings to prevent the spread of the virus. As the Influenza waned, evidence of the positive impact of windows opening, even during winter months,

ISOLATING WARD.
VIEW FROM NORTHEAST.

Figure 3.15
Above: The 1873 Johns Hopkins Hospital Isolating Ward in Baltimore, Maryland, US. Notice the many ventilation chimneys. Source: Wellcome Library, London, Wellcome Images.

remained. Oversized more powerful radiators began to be specified and placed under windows in American homes, schools and other buildings such that icy winter breezes passing through cracked windows would not undermine indoor thermal comfort. As time passed and the memory of the Influenza pandemic receded, people began to shut windows again. The heat from these oversized radiators was so intense that a new heat-reflecting silver paint was used to keep rooms from becoming uncomfortably warm.[37,38]

Covid-19 and the Deadly Impact of Sealed, Un-Ventilated Spaces

The Covid-19 pandemic has served as a grim reminder of the consequences of ignoring conventional wisdom about the value of natural ventilation. As learnings from of the 1918 Influenza pandemic were mostly forgotten, the value of opened windows also faded. In its place, sealed spaces and HVAC, principal elements of curtain wall International Style modernism, have largely overtaken the operable window. At the time of writing this book, the author's office was located in a converted public middle school building on a college campus, constructed in 1934. Inside the building, second floor rooms with 8 feet tall windows could be opened as much as 4 feet without the concern of falling out (this is obviously a concern on higher floors).[39] As was the case of the author's window, sometimes opening is obstructed by a window-mounted air-conditioning unit. Nevertheless, the ability to open a window for "gas exchange" or the introduction of fresh air and the expulsion of stale air is a clear benefit that had been covered widely during the Covid-19 pandemic.[40]

Nearby the author's building, a similar structure with operable windows and natural ventilation features was demolished to make way for a new student center and dormitory. The design of this new innovation center (mentioned earlier), eschewed operable windows in favor of sealed glass and a powerful HVAC system. Together, this decision requires powered fans to provide fresh air. Ironically, the building in question was under construction just as the Covid-19 virus appeared. What once seemed like a reasonable, conventional way of providing fresh air indoors was now suddenly being questioned by public and international health officials. Overnight, recirculated air was deemed dangerous as virus-filled human aerosols produced by infected people could be transmitted from one room to another—or within a room that is mostly sealed. A now famous case of virus transmission from recirculating air in a windowless Guangzhou, China restaurant confirmed the dangers of poor ventilation.[41] In contrast, the age-old practice of opening windows to the outside to obtain fresh air became a standard health department recommendation for all enclosed businesses.[42]

Using openable windows to increase air flow was also a likely impetus behind the design of early examples of public transportation. Streetcars, subways and trains from the turn of the 20th century through the Second World War were often equipped with openable windows for cooling and to promote air quality. In some cases, streetcars were designed to be open air and trains could open their front windows to allow vigorous breezes through the car. But even smaller windows on American school buses can promote a tremendous amount of fresh air gas exchange. Whereas exchanging air in a room six times an hour is now considered safe, one study demonstrated that open windows in a public school bus could exchange the indoor air as much as 20–40 times per hour.[43]

Today, in contrast, much of public transportation is enclosed for air-conditioned climate control. In New York City's metropolitan transit system, most cars have small windows that can be opened, but for the most part, air is recirculated by the air-conditioning system. More recent signs in these cars proclaim "Air conditioned car, please close windows." Regional trains are even worse. Because of the speed at which they travel, contemporary intercity train cars are sealed shut, relying entirely on recirculated air to ventilate and cool the interior. Keeping safety in mind, train and public transit designers could consider window retrofits for existing cars and partially opening windows or vents for future train sets.

Electric and Whole House Fans

Fans, like open windows, can help circulate fresh air and cool people simultaneously. Philip Diehl is sometimes credited with inventing the first electric fan after patenting such a device in 1889.[44] While some evidence suggests that the first rotary fan was developed in China using human propulsion to separate grain,[45] electric fans for cooling became popular around 1900 as homes began to be electrified. Models designed and manufactured by General Electric, AEG and others provided much welcomed relief from summer heat by creating an artificial breeze. Still, electric fans use far less energy than air conditioners.[46] Today, oscillating and ceiling variants continue to be useful, except they are generally most effective when people are in the path of their air flow. Turning on a fan will not appreciably reduce the temperature of a room, but it will cool a person by blowing wind over their skin to promote evaporative cooling (perspiration).[47] They can also be used in conjunction with air conditioning to better circulate cool air. During the energy crises of the 1970s their popularity returned in the US as consumers sought ways to reduce their energy costs.

However, decades before, the now obscure *whole house fan* was introduced providing a more integrated system of mechanical breeze in a home. Specifically, whole house fans provide additional cooling performance by expelling warmer air from a building through an attic vent and by drawing cooler air into lower floors. The Chelsea Fan company, for example, once produced a system in the mid 20th century that simultaneously opened a top floor ceiling louver into the attic while starting an extra-large attic fan to draw warm air out from an entire house. Similar to the stack effect cooling techniques of wind catchers and Kéré Architecture's Startup Lions campus project, whole house fan systems still work actively while consuming much less energy than air conditioners. Some systems survive, albeit in much fewer numbers, and some households don't even use them if they are still present and operational. Even though they have even been promoted by the US Department of Energy just as ceiling fans have been, their usefulness has been largely overlooked or replaced with air conditioning.[48] While many homes today have smaller attic fans, home owners often don't know how to use them to maximum effect. Today, there are a new generation of smartphone devices to open windows and support the stack-effect cooling process.[49] There is certainly ample opportunity to design houses today with integrated whole house fan systems, using quieter, perhaps even *Dyson* fan technology. For the convenience-minded, these devices could be further automated to coordinate the attic fan, opening louvers *and* basement or lower floor windows to draw cooler air inside. Even with the additional energy expenditure of such automation, the system would surely consume far less energy than air conditioning.

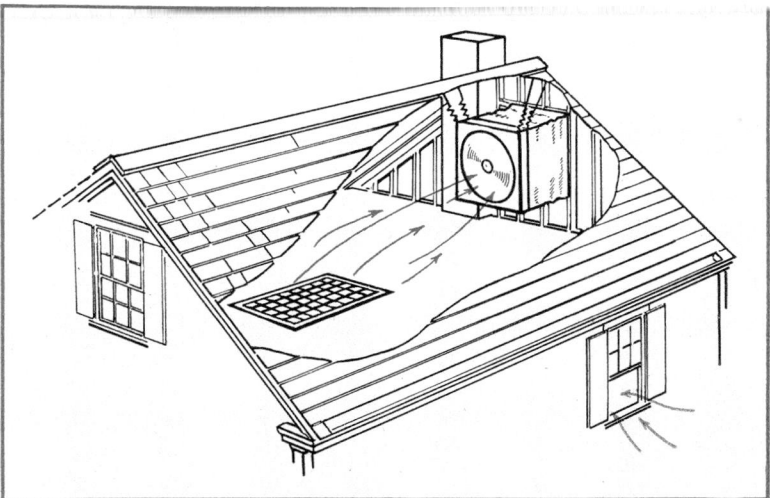

This diagram shows air movements when fan is placed close to an outside wall and connected direct to louvers by a duct. Fan is not connected directly to ceiling grille. The attic serves as a plenum chamber. Fan unit is suspended from rafters by springs.

Figure 3.16
A Chelsea whole house fan system promotional diagram dating from before the 1950s. Chelsea fan systems and their celling louvers were activated by a switch but required home owners to open and close their windows to allow cool air in after sun down. Source: Chelsea Fan Company.

Mechanized Evaporative Cooling in Homes, Office Buildings and Cars

Despite the efficacy of wind catchers paired with qanats, passive evaporative cooling technology did not infiltrate Western culture extensively until around 1900. Roughly at the same time, Willis Carrier invented the first operable air-conditioning unit, engineers began experimenting with the same basic principles of wind catcher and qanat systems except using electric-powered fans. John Zellweger, an inventor from St. Louis, Missouri developed and patented an early evaporative cooler in 1906.[50] Titled an "air filter and cooler," the technology developed through the mid-century and was even installed centrally in houses and as an accessory for automobiles after the Second World War. By the 1950s, as air-conditioning units became smaller and more affordable, they overtook evaporative coolers in the marketplace. Yet today, evaporative coolers have the distinct advantage of lower up-front costs and much lower operating costs compared with air conditioning.[51] The disadvantage is that they are mostly effective in drier climates like the American southwest, not the humid climates of the eastern US, Africa, Europe and Asia. Nevertheless, there is a solid rationale for reintroducing evaporative coolers in many parts of the world today in lieu of air conditioning, including potentially in the mass transportation sector, where air conditioning consumes large amounts of energy.[52] Significant research is now also being conducted in ways to reintroduce evaporative cooling in architecture, using complex 3d printed clay structures to help disperse cooler air in local spaces and environments.[53]

Figure 3.17
Above left: A late 1940s American car with an optional window-mounted evaporative cooler. These amenities were offered optionally in the American southwest but were later replaced by air-conditioning systems which work in all climates despite creating more drag on a car's engine and fuel economy. Also note the windshield visor: these were popular in the 1940s and '50s to help reduce glare and lower internal temperatures from sun load. Source: Doug Coldwell.

Figure 3.18
Above right: A concept evaporative cooler 3d printed in clay and based on the same principles as Zeer pots. When filled with compacted sand and water, the water evaporates through the porous wall, thereby cooling the air. The vessel could potentially be displayed as a center piece on a table in a room. Source: Saeed Sakhdari.

Passive Heating: Trombe Walls and Thermal Mass

A Trombe wall is a passive architectural feature that can be used to bring solar energy inside. Named after French engineer Félix Trombe, who was responsible for bringing attention to this type of heating system in the 1960s, the origin of the idea goes back much further. In 1881, for example, a thermal-mass wall was patented by Edward Morse (Figure 3.19). In the US, interest in Trombe walls emerged in the 1970s, aided by researchers at Los Alamos National Laboratory in New Mexico.[54] A Trombe wall consists of a masonry slab in the interior of a building facing the equator next to a glass window positioned several inches away. The glass allows sunlight to come into the building and heat the wall. In a temperate climate, a roof overhang or shade trees can be used to shade the wall in the summer when the angle of the sun is higher and when heating is not desired, functioning much like Brise Soleil features that were discussed earlier. Operable vents can be opened to allow warm air to enter the space in the winter and closed to keep warm air out of the space in the summer. Such features could be especially useful now, integrated into new buildings, as a part of broader Passive House, LEED and NetZero housing initiatives.

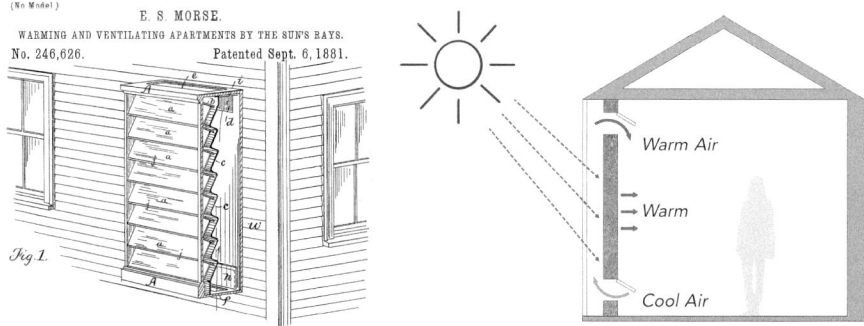

Figure 3.19
Above left: An 1881 Trombe wall window fitting invention by Edward Morse. Note from the diagram how cooler air below is introduced into the chamber between the window and masonry fitting, warming in the process and then reintroducing the passively heated air back into the house. Source: USPTO.

Figure 3.20
Above right: A schematic illustration of a Trombe wall. Source: Brook Kennedy.

Seasonal Temperature Control: Shade at Cliff Palace Mesa Verde

Cliff Palace in southwest Colorado is the largest of several structures built into rocky mesa cliffs by the Puebloan peoples sometime between roughly 600 and 1300AD. After this time, the structures were abandoned. Centuries afterward, they were discovered by European colonists in the late 19th century. In 1906, the structures were preserved by the US National

Figure 3.21
Cliff Palace, constructed between 600AD and 1300AD and later abandoned. Cliff Palace is designed to maximize heat gain during the winter while the overhanging rock provides shade during the summer. Source: Greg Tally (CC BY-SA 3.0).

Park Service and remain protected today by the US federal government and UNESCO, which designated Cliff Palace and companion structures a World Heritage Site in 1978. Numerous studies have emphasized how the architecture of Cliff Palace takes advantage of passive solar energy.[55] First of all, the structures face south, giving them maximum sun exposure. In the wintertime, the sun's angle is low enough to provide some thermal gain for the buildings. Conversely, during the hot dry summers, the overhanging rock cliff provides shade to maintain a cooler, more comfortable temperature indoors. Furthermore, earthen construction and the surrounding cliff rock provides insulating thermal mass to regulate the temperature in the dwellings throughout the year. Simple and effective, surely these techniques could be more widely adopted in new buildings today.

The Enduring Value of Shade and Trees

Shade has long provided cooling relief from summer heat. Whether natural or artificially created, shade has played a vital role in reducing solar heat gain through windows and making outdoor spaces usable during the warmest months. Covered porches (and related architectural features), whether those incorporated in buildings constructed over two millennia ago or the craftsman bungalows constructed in great numbers throughout the first half of the 20th century, provide some passive cooling effects. By the post-Second World War period, the functioning porch largely disappeared from residential architecture, in part as air conditioning became more affordable. Trees and landscaping have also played a vital role in keeping buildings cool and directing breezes strategically around dwellings. Shade trees, for example, have long been used to cool temperatures in buildings during the summer, while allowing sun to pass through to warm walls or penetrate windows after leaves have fallen during winter months.[56,57] Trees, landscaping and house placement in relation to hills can also be used to manage persistent winds which can promote breezes, and to cool homes or protect homes from windchill. For winter heating retention, evergreen trees can be beneficial for temperature management because of their year-round density.[58]

Figure 3.22
Above left: The Gamble House in Pasadena, California by architects Greene and Greene. The Gamble House is credited for popularizing the widely produced California Bungalow style with its deep overhanging eaves, porches, shutters and limits on window sizes to manage heat in the southern California climate. The House adapted to other climates. Source: Cullen328 (CC BY-SA 3.0).

Figure 3.23
Above right: An image from an October, 1994 report published by the US National Renewable Energy Lab promotes the significant cooling value of planting and using shade trees. Source: National Renewable Energy Laboratory (NREL).

House Zero and The Passivhaus Standard: Blending the Best of the Past and Future for Energy Saving and Carbon Neutrality

The Center for Green Buildings and Cities at the Harvard University Graduate School of Design partnering with Swedish Architecture firm Snøhetta teamed together in 2016 to design and build a zero net energy consuming home using alternative approaches to heating and cooling. Of special interest, the house was retrofitted from a 1920s single-family house in Cambridge, Massachusetts to demonstrate how existing building stock, which contributes enormously to overall energy consumption, could be modified for radically improved energy efficiency. As an ongoing research experiment, thousands of sensors are placed in the building to measure heat flows to further inform and augment energy reduction. Intended as a living laboratory of energy-efficient building, the structure stands as a viable proof of concept for the over 110 million homes in the US that consume as much as 21% of all energy used in the nation.[59] Many of the house's features draw from previously discussed alternative technologies blended with emerging energy saving techniques, from Solar Chimneys, geothermal heating, passive shading and many more to achieve this result. In contrast, the Passivhaus standard relies more on new technology to radically reduce carbon emissions and energy consumption in buildings. Originated in the 1980s with the foundation of the Passivhaus Institute in Darmstadt Germany, variations of the standard have been established in Sweden, Switzerland, Canada and the US. Overlapping with House Zero's approach, the Passivhaus standard relies chiefly on modern insulation and triple panel windows to seal temperatures indoors. At the same time, both the Passivhaus and House Zero use some form of passive geothermal energy to help keep houses cool without overreliance on air conditioning. House Zero also uses shade, stack effect ventilation and automatically opening windows to induce fresh air flow and additional cooling. While neither house standard might work in every global climate, they offer solid examples of the enduring relevance of alternative and traditional knowledge in defining how to build more responsibly today and in the future.

Figure 3.24
Above left: House Zero, Cambridge, MA, US. Notice the shade promoting window frames, the solar chimney for promoting air flow and automatically opening windows. Source: Michael Grimm, Photographer, Snøhetta.

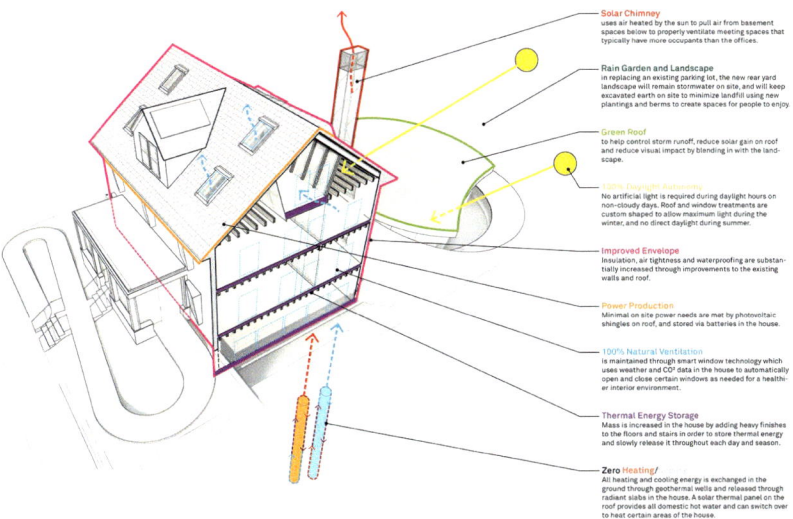

Solar Chimney
uses air heated by the sun to pull air from basement spaces below to properly ventilate meeting spaces that typically have more occupants than the offices.

Rain Garden and Landscape
in replacing an existing parking lot, the new rear yard landscape will remain stormwater on site, and will keep excavated earth on site to minimize landfill using new plantings and berms to create spaces for people to enjoy.

Green Roof
to help control storm runoff, reduce solar gain on roof and reduce visual impact by blending in with the landscape.

100% Daylight Autonomy
No artificial light is required during daylight hours on non-cloudy days. Roof and window treatments are custom shaped to allow maximum light during the winter, and no direct daylight during summer.

Improved Envelope
Insulation, air tightness and waterproofing are substantially increased through improvements to the existing walls and roof.

Power Production
Minimal on site power needs are met by photovoltaic shingles on roof, and stored via batteries in the house.

100% Natural Ventilation
is maintained through smart window technology which uses weather and CO_2 data in the house to automatically open and close certain windows as needed for a healthier interior environment.

Thermal Energy Storage
Mass is increased in the house by adding heavy finishes to the floors and stairs in order to store thermal energy and slowly release it throughout each day and season.

Zero Heating/
All heating and cooling energy is exchanged in the ground through geothermal wells and released through radiant slabs in the house. A solar thermal panel on the roof provides all domestic hot water and can switch over to heat certain areas of the house.

Figure 3.25
Above right: Schematic diagram of House Zero. Source: Snøhetta.

Dressing for the (Thermal) Occasion

As we have discussed in this chapter, heating and cooling spaces account for a significant and growing portion of human energy use, especially in industrialized nations. Yet before the centralized systems became commonplace around the world, individual dress behaviors and designs provided viable alternatives for controlling thermal comfort. As technology writer Kris De Decker summarizes, society once prioritized, "heat[ing] the body, not the room."[60] Diverse cultures worldwide provide examples of dress and personal accessories that are suited to the local climate, from societies around the equator and the Middle East which use traditional garments to adapt to the heat, whereas other cultures located closer to the most polar tips of the globe have adapted equally to the cold. More recently, formal business dress codes, which can make you feel warmer, were relaxed as part of the "Cool Biz" Campaign in Japan. In the summer of 2005, the Japanese government inaugurated the program to help reduce energy consumption from air conditioning. Devised by then Minister Yuriko Koike, the campaign is composed of two parts: first thermostats are set to 28 degrees Celsius (82 degrees Fahrenheit); and second, government workers are allowed to dress casually from June to September. The dress code consists of allowing men to come to work without ties and blazers and to wear short sleeve shirts. Many would dress up more formally on their commutes to culturally uphold a formal outer appearance in public only to dress down upon arrival in the office.[61] Given the disconnect between current standards of fashion formality and summer temperatures, what could be learned from other dress traditions? What if the design of more formal attire could evolve to balance these objectives? Traditional Thobes manage to achieve this objective. Still worn today in the Middle East and Africa to address the effects of heat and intense sun exposure, they balance tradition, formality and comfort. Loose fitting and often lighter in color, they reflect sunlight and enable those wearing them to sweat, and for the sweat to evaporate. By covering the skin, they limit sun exposure, both to prevent skin damage and to avoid heat exhaustion.

Other overlooked types of clothing have attempted to balance these demands as well. Brimmed hats were an essential part of Western formal dress until the second half of the 20th century and are still used electively during informal settings when sun exposure is a concern. Another useful traditional tool is the parasol—a cooling, sun-protecting variant of the umbrella. At one time, parasols evolved into a necessary fashion statement, seen in many impressionist and pointillist paintings from late 19th-century France. However, by the 20th century, the parasol mostly disappeared from use. Today umbrellas are still sometimes adapted for this purpose. Surely the parasol could be reimagined and redesigned with contemporary materials to represent present fashion and appearance standards—and to serve as a contemporary passive cooling accessory.

Notes

1 International Energy Administration (2019) *Global air conditioner stock, 1990–2050* [online]. Available at: www.iea.org/data-and-statistics/charts/global-air-conditioner-stock-1990-2050 (Accessed: August 8, 2022).

2 International Energy Administration (2018) *Percentage of households equipped with AC in selected countries, 2018 (Updated Nov 16, 2019)* [online]. Available at: www.iea.org/data-and-statistics/charts/percentage-of-households-equiped-with-ac-in-selected-countries-2018 (Accessed: August 4, 2022).

3 US Energy Information Agency (2015) *Residential Energy Consumption Survey* [online]. Available at: www.eia.gov/todayinenergy/detail.php?id=36412&src=‹%20Consumption%20%20%20%20%20%20Residential%20Energy%20Consumption%20Survey%20(RECS)-b7 (Accessed: August 4, 2022). This report estimates that AC accounts for as much as 17% of home electricity use.

4 US Energy Information Agency (2018) *Air conditioning accounts for about 12% of U.S. home energy expenditures* [online]. Available at: www.eia.gov/todayinenergy/detail.php?id=36692 (Accessed: July 29, 2022).

5 US Energy Information Agency (2021) *Electricity explained: Electricity in the United States* [online]. Available at: www.eia.gov/energyexplained/electricity/electricity-in-the-us.php (Accessed: August 4, 2022).

6 International Energy Administration (2018) 'The future of cooling opportunities for energy-efficient air conditioning. Global air conditioner stock, 1990–2050' [online]. Available at: www.iea.org/data-and-statistics/charts/global-air-conditioner-stock-1990-2050 (Accessed: August 4, 2022).

7 Roaf, S. (2008) 'The traditional technology trap (2): More lessons from the windcatchers of Yazd,' *PLEA 2008–25th Conference on Passive and Low Energy Architecture*, Dublin, 22 to 24 October 2008.

8 See Chapter 4, 'The cooling effect: Air-conditioning,' in Al, S. (2022) *Super tall: How the world's tallest buildings are reshaping our cities and our lives*. New York: W.W. Norton. Al specifically talks about how the skyscraper's development was closely linked with air conditioning.

9 Basile, S. (2014) *Cool: How air conditioning changed everything*. New York: Fordham University Press.

10 Ackermann, M.E. (2010) *Cool comfort: America's romance with air-conditioning*. Washington: Smithsonian Institute Press.

11 Denzer, A. (2013) The solar house: Pioneering sustainable design. New York: Rizzoli.

12 McCoy, E. (1977) *Case study houses: 1945–1962*. Los Angeles: Hennessey & Ingalls.

13 Sisson, P. (2017) 'How air conditioning shaped modern architecture—and changed our climate,' *Curbed*, May 9 [online]. Available at: https://archive.curbed.com/2017/5/9/15583550/air-conditioning-architecture-skyscraper-wright-lever-house (Accessed: August 3, 2022).

14 Wren, C.S. (1999) 'International symbol of neglect; U. N. Building, unimproved in 50 years, shows its age,' *The New York Times*, October 24 [online]. Available at: www.nytimes.com/1999/10/24/nyregion/international-symbol-neglect-u-n-building-unimproved-50-years-shows-its-age.html (Accessed: August 3, 2022).

15 Short, C.A. (2017) 'Back to the future of skyscraper design' [online]. Available at: www.cam.ac.uk/research/news/back-to-the-future-of-skyscraper-design (Accessed: October 5, 2022).

16 Passipedia. (2020) 'What is Passive House?' [online]. Available at: https://passipedia.org/basics/what_is_a_passive_house (Accessed: August 12, 2022).

17 U.S. Department of Energy (n.d.) *Why use energy-efficient window attachments?* [online]. Available at: www.energy.gov/energysaver/energy-efficient-window-attachments (Accessed: August 8, 2022).

18 Barber, D. (2020) *Modern architecture and climate: Design before air conditioning.* Princeton: Princeton Architectural Press.

19 Kassir, R.M. (2015) 'Passive downdraught evaporative cooling wind-towers: A case study using simulation with field- corroborated results,' *Building Services Engineering Research and Technology*, 37(1) pp103–120.

20 Ghadiri, M. (2011) 'The effect of tower height in square plan wind catcher on its thermal behavior,' *Australian Journal of Basic and Applied Sciences*, 5(9) pp381–385.

21 Soelberg, C. and Rich, J. (2014) 'Sustainable construction methods using ancient BAD GIR (Wind Catcher) technology,' *Construction Research Congress* [online]. Available at: https://doi.org/10.1061/9780784413517.161 (Accessed: August 8, 2022).

22 Malone, A. (2012) 'The windcatcher house,' *Architectural Record* [online]. Available at: www.architecturalrecord.com/articles/6495-the-windcatcher-house (Accessed: August 8, 2022).

23 Rosen, R.J. (2011) 'Keepin' it cool: How the air conditioner made modern America,' *The Atlantic*. Available at: www.theatlantic.com/technology/archive/2011/07/keepin-it-cool-how-the-air-conditioner-made-modern-america/241892/ (Accessed: August 8, 2022).

24 Spencer, I. (2007) 'Dogtrot house,' *Architectural Record*. Available at: www.architecturalrecord.com/articles/8570-dogtrot-house (Accessed: August 8, 2022).

25 'Vindauga,' in Zoëga, G.T. (1910) *A concise dictionary of Old Icelandic.* Oxford: Clarendon Press.

26 Environmental Protection Agency (1991) *Indoor Air Facts No. 4 (revised) Sick Building Syndrome* [online]. Available at: www.epa.gov/sites/default/files/2014-08/documents/sick_building_factsheet.pdf (Accessed: August 8, 2022).

27 Bauer, A., Möller, S., Gill, B. and Schröder, F. (2020) 'When energy efficiency goes out the window: How highly insulated buildings contribute to energy-intensive ventilation practices in Germany,' *Energy Research & Social Science*, 72.

28 Martinot, E. (1997) *Investments to improve the energy efficiency of existing residential buildings in countries of the Former Soviet Union.* 1st print edn. Washington, DC: World Bank.

29 Fowler, D. et al. (2020) 'A chronology of global air quality,' *Philosophical Transactions of the Royal Society*, September 28 [online]. https://doi.org/10.1098/rsta.2019.0314 (Accessed: August 8, 2022).

30 Jasarevic, T., Thomas, G. and Osseiran, N. (2014). *7 million premature deaths annually linked to air pollution* [online]. Available at: www.who.int/news/item/25-03-2014-7-million-premature-deaths-annually-linked-to-air-pollution (Accessed: August 8, 2022).

31 Fisher, T. (2010) 'Frederick Law Olmsted and the campaign for public health,' *Places* [online]. Available at: https://placesjournal.org/article/frederick-law-olmsted-and-the-campaign (Accessed: August 8, 2022).

32 Plunz, R. (1990) *A history of housing in New York City.* New York: Columbia University Press.

33 Howard, E. (1898) *Garden cities of to-morrow.* London: Swan Sonnenschein & Co.

34 Kiechle, M.A. (2021) 'Revisiting a 19th century medical idea could help address covid-19,' *The Washington Post* [online]. Available at: www.washingtonpost.com/outlook/2021/04/21/revisiting-19th-century-medical-idea-could-help-address-covid-19/ (Accessed: August 8, 2022).

35 Short, C.A. (2017) *The recovery of natural environments in architecture: Air, comfort and climate.* London: Routledge.

36 Ishida, Aki. (2020) 'Imprints of an invisible virus: How airborne diseases change cities,' *The Plan Journal*, 5(2). Available at: doi 10.15274/tpj.2020.05.02.3 (Accessed: August 8, 2022).

37 Sisson, P. (2020) 'Your old radiator is a pandemic-fighting weapon: Turn-of-the-century faith in ventilation to combat disease pushed engineers to design steam heating systems that still overheat apartments today,' *Citylab*, August 5 [online]. Available at: www.bloomberg.com/news/articles/2020-08-05/the-curious-history-of-steam-heat-and-pandemics (Accessed: August 8, 2022).

38 Hall, D. (2021) 'Radiant point,' *99% Invisible* [online]. Available at: https://99percentinvisible.org/episode/mini-stories-volume-11/ (Accessed: August 8, 2022).

39 Window opening is not as practical in higher floors for safety and other reasons. For the tallest of skyscrapers, keeping windows closed is a matter of necessity; high winds entering upper floors slam doors and blow papers off of desks. See AI (2020).

40 Landsverk, G. (2020) 'A gym trainer exposed 50 athletes to COVID-19, but no one got sick — because one member is a ventilation expert who redesigned the room's layout,' *Insider*, November, 11. Available at: www.insider.com/how-gym-prevented-outbreak-after-coach-got-covid-19-2020-11 (Accessed: August 8, 2022).

41 Jianyun Lu et al. (2020) 'COVID-19 outbreak associated with air conditioning in restaurant, Guangzhou, China, 2020,' *Emerging Infectious Diseases*, 26(7). Available at: wwwnc.cdc.gov/eid/article/26/7/20-0764_article (Accessed: August 8, 2022).

42 Hughes, M. (2020) 'A gym trainer exposed 50 athletes to Covid-19, but no one else got sick because of a ventilation redesign,' *CNN*, November 9 [online]. Available at: www.cnn.com/2020/11/19/us/gym-ventilation-covid-trnd/index.html (Accessed: August 8, 2022).

43 Abulhassan, Y. and Davis, G.A. (2021) 'Considerations for the transportation of school aged children amid the Coronavirus pandemic,' *Transportation Research Interdisciplinary Perspectives*, 9 [online]. Available at: www.sciencedirect.com/science/article/pii/S2590198220302013 (Accessed: August 8, 2022).

44 Diehl, P. (1889) 'Electric motor fan.' Available at: https://worldwide.espacenet.com/patent/search/family/002483687/publication/US414758A?q=pn%3DUS414758 (Accessed: August 8, 2022).

45 Needham, J. (1965) *Science and civilization in China, Vol. IV: Physics and physical technology.* Cambridge: Cambridge University Press.

46 Southface Energy Institute and the Oak Ridge National Labs (1999) 'Whole house fan: How to install and use a whole house fan,' *US Department of Energy* [online]. Available at: www.nrel.gov/docs/fy99osti/26291.pdf (Accessed: August 8, 2022).

47 De Decker, K. (2015) 'Restoring the old way of warming: Heating people, not places,' *Lowtech Magazine*. Available at: www.lowtechmagazine.com/2015/02/heating-people-not-spaces.html (Accessed: August 8, 2022). Alternative technology author Kris De Decker writes persuasively about the topic of localized thermal comfort and energy use. Speaking specifically about heating he makes the following insightful observation: "These days, we provide thermal comfort in winter by heating the entire volume of air in a room or building. In earlier times, our forebear's concept of heating was more localized: heating people, not places." Adapting this principal to cooling, electric fans too can provide individualized cooling, using far less energy than air conditioning an entire house. De Decker also stresses, importantly, the need for past technological principles to be updated using present scientific knowledge and materials to advance their utility for today.

48 Southface Energy Institute and the Oak Ridge National Labs (1999).

49 At the time of writing this book, several startup companies including Fenestra were developing automatic window-opening devices powered by solar energy. These could be combined with whole house fan systems to cool homes, using less energy than air conditioning.

50 Zellweger, J. (1906) *Air cooler and filter* [online]. Available at: https://patentimages.storage.googleapis.com/e0/63/2d/41c28f5a4752a1/US838602.pdf

51 U.S. Department of Energy (n.d.) *Evaporative coolers* [online]. Available at: www.energy.gov/energysaver/home-cooling-systems/evaporative-coolers (Accessed: August 8, 2022).

52 IEA (2019) *Cooling on the move. The future of air conditioning in vehicles*. Available at: https://iea.blob.core.windows.net/assets/425608f4-4368-42ca-8faa-568d241ce7ea/Cooling_on_the_Move.pdf

53 Gan, A.W.J., Guida, G., Kim, D., Shah, D., Youn, H. and Seibold, Z. (2022) 'Modulo Continuo: 5-Axis ceramic additive manufacturing applications for evaporative cooling facades modules,' in Pak, B., Wurzer, G. and Stouffs, R. (eds.) *Co-creating the future: Inclusion in and through design – proceedings of the 40th conference on education and research in Computer Aided Architectural Design in Europe* (eCAADe 2022), Ghent, 13–16 September 2022, pp. 47–55.

54 Childs, K.W., Courville, G.E. and Bales, E.L. (1983) 'Thermal mass assessment: An explanation of the mechanisms by which building mass influences heating and cooling energy requirements,' *U.S. Department of Energy Office of Scientific and Technical Information* [online], report number: ORNL/CON-97 ON: DE84000654. Available at: www.osti.gov/biblio/5788833 (Accessed: August 8, 2022).

55 Lea, K. (2010) *Mesa Verde cliff dwellings* [online]. Available at: https://greenpassivesolar.com/2010/04/mesa-verde-cliff-dwellings/ (Accessed: August 8, 2022).

56 Simpson, J.R. and McPherson, E.G. (1998) 'Simulation of tree shade impacts on residential energy use for space conditioning in Sacramento,' *Atmospheric Environment*, 32(1), pp69–74. https://doi.org/10.1016/S1352-2310(97)00181-7

57 Pandita, R. and Laband, D.N. (2010) 'Energy savings from tree shade,' *Ecological Economics*, 69(6), pp1324–1329.

58 Lechner, N. (2015) *Heating cooling lighting: Sustainable methods for architects*. 4th edn. Hoboken: John Wiley & Sons.

59 U.S. Department of Energy's Building Technologies Office (2020) *National residential energy facts* [online]. Available at: https://rpsc.energy.gov/energy-data-facts (Accessed: August 9, 2022).

60 De Decker, K. (2015) 'Heating people, not places: How to keep warm in a cool house,' *Resilience*, March 17 [online]. Available at: www.resilience.org/stories/2015-03-17/heating-people-not-places-how-to-keep-warm-in-a-cool-house/ (Accessed: August 8, 2022).

61 Environmental and Energy Study Institute (2015) *The Japanese cool biz campaign: Increasing comfort in the workplace* [online]. Available at: www.eesi.org/articles/view/the-japanese-cool-biz-campaign-increasing-comfort-in-the-workplace (Accessed: August 8, 2022).

Chapter 4

Daylighting vs. Electric Lighting

Natural lighting is widely valued in architecture and has been for ages. At the same time, tradeoffs between admitting natural light and limiting solar heat gain requires a delicate balance. For ages, walls made of masonry, wood and other materials protected inhabitants from the elements, especially in colder climates, but deprived indoor spaces of sufficient light to see during the daytime. Windows allowing sunlight indoors developed slowly, so candles, oil lamps and other flame-based light sources made it possible to do basic tasks indoors after dark. By the 19th century, iron and steel construction along with glass-making technology enabled walls and sometimes roofs and floors to be opened up to the outdoors

Figure 4.1
The Pantheon, c. 126AD, painted by Giovanni Paolo Panini in 1747. The Oculus at the apex of the dome allows natural light (and rain) into the building. Source: Wmpearl (CC0 1.0).

 DOI: 10.4324/9780367814304-5

and natural light. This chapter will focus on ways in which natural light has been manipulated by design before and coincident with the development of artificial and electric light. Today, despite new efforts to advance daylighting, electric lighting remains a dominant form of interior illumination. Certainly, electric lighting technology has made significant leaps in efficiency from the incandescent bulb to the Light Emitting Diode (LED). At present, lighting accounts for roughly 6% of energy use overall in the US.[1] Notwithstanding, this chapter will summarize alternative, passive indoor systems for daylighting, which use little, if any, energy at all. The role of windows, their size, light wells, floor mounted glass block and other passive technologies like prism glass will be considered in contrast to incrementally improving electric lighting. Often electric lights are left on during the day and night even when unused. Newer sensor activated lighting systems, already widely used outside the US, deserve mention but will not be covered here. Instead, fiber optic daylighting, heliostats and other natural light-channeling solutions will be considered.

The benefits of natural light are numerous, beyond the practicalities of being able to see and perform tasks indoors. Significant literature has focused on the positive psychological and physiological effect of exposure to natural light, ranging from regulation of sleep patterns, vitamin D retention and general mental health.[2] On the other hand, excessive lighting, often referred to as light pollution, has also been questioned.[3] Among its critics is noted author of *In Praise of Shadows*, Jun'ichirō Tanizaki, who has written about Japanese cultural associations with lighting as it relates to aesthetics and indoor ambience.[4]

Lighting and Industrial Productivity

During and after the industrial revolution, factory and business owners desired to increase indoor lighting to promote worker productivity and thereby extend the output of their work day. For centuries, candle light and oil lamps were the main means of illuminating indoor spaces, especially in rooms without windows and after dark. Then in the 18th and 19th centuries, gas lighting emerged as a promising technology, despite the noxious fumes associated with the combustion of gaseous fuels. As early as 1802, the Boulton and Watt Soho Foundry in the UK was illuminated by roughly 2,600 gas lights to permit night time operations.[5] While incandescent mantles later amplified gas lighting by glowing brightly when heated by a gas flame (beginning around 1885), electric incandescent lighting would ultimately prevail a few decades later. In a few rare cases, some towns and cities in the US, UK and around the world have preserved their outdoor gas lights for atmospheric and nostalgic effect.[6]

Several forms of electric lighting began to appear beginning in the 18th century through the late 19th century, but it was not until Thomas Edison that the incandescent bulb reached a point of scalable reliability around 1879.[7] Furthermore, Edison developed an early electric grid, with switches, enabling customers to receive electricity from a central provider in a small perimeter of lower Manhattan. Electric street lamps were also installed on Pearl Street in lower Manhattan in 1882.[8] At the same time, passive design techniques aimed at increasing natural light indoors paralleled the development of electric light. New building technologies, particularly structural iron and steel, enabled architects to begin opening walls and façades to allow for larger glass window surfaces. For example, at London's 1851 Great Exhibition of the Works of Industry of All Nations, Joseph Paxton's Crystal Palace Hall showcased a new frontier of glazed wall (and roof) construction. Built of a cast iron frame and clad in newer, less costly plate glass windows, natural light flooded the interior of the building during the five and a half months the exhibit remained open. This type of construction inspired new designs for building features aimed at bringing natural daylight indoors. On a smaller scale, glass light courts and skylights began to appear, especially in top floors of buildings and in stairwells. By the late 1860s, cast iron ornamental façades with large windows (as seen in the Soho neighborhood of New York City) enabled ample daylight indoors for industrial activities. Opening walls to glass even further continued with examples like the Reliance Building in Chicago (Burnham and Root, 1891), La Maison de Verre (Pierre Chareau and Bernard Bijvoet, 1928) and the Bauhaus (Walter Gropius, 1928). Eventually, curtain-wall structures ranging from the Farnsworth House (Ludwig Mies Van der Rohe, 1945) and Lever House (Gordon Bunschaft and Natalie De Blois, 1950) emerged. Today, many of the tallest skyscrapers around the world continue to be designed this way including One Vanderbilt (Kohn Pedersen Fox, 2020) to maximize natural light.[9] The development of such wall construction greatly improved natural lighting conditions indoors for workers, yet indoor work spaces still relied heavily on artificial light and increasingly on air conditioning.

Figure 4.2
Above left: An exterior view of Joseph Paxton's 1851 all glass and cast iron Crystal Palace in London. Source: Read & Co. Engravers & Printers.

Figure 4.3
Above right: Pierre Chareau's and Bernard Bijvoet's 1928 Maison de Verre in Paris with a façade made mostly of glass block. Source: August Fischer (CC BY-ND 2.0).

Glass Block and Prism Glass

Building upon designs used in ships in the late 18th and early 19th centuries, cast glass blocks and prism glass began to be adapted to architecture to direct natural light indoors and underground. As Dietrich Neumann's article *The Century's Triumph in Lighting* thoroughly investigates, manufacturers like the Luxfer Prism company perfected methods of casting glass tile to provide natural indoor illumination during the 1890s without electricity or fuels. Using optic principles developed by Augustin-Jean Fresnel in the early 19th century, prism glass redirects natural light into the deep interiors of office and industrial buildings. On the lower floor of buildings, prism glass can redirect light deep into a space. Often, prism glass was hung above ground-floor plate glass windows in transoms or mounted as plates on hinges above windows where they could be adjusted during the day to optimize the direction of illumination. So called "Sidewalk Lights" and "Vault Lights" were used to redirect light underground.

Figure 4.4
Above left: An advertisement for the American Luxfer Prism Company in Chicago, US. Source: American Luxfer Prism Co.

Figure 4.5
Above right: A design patent for a prism light designed by architect Frank Lloyd Wright. Source: USPTO.

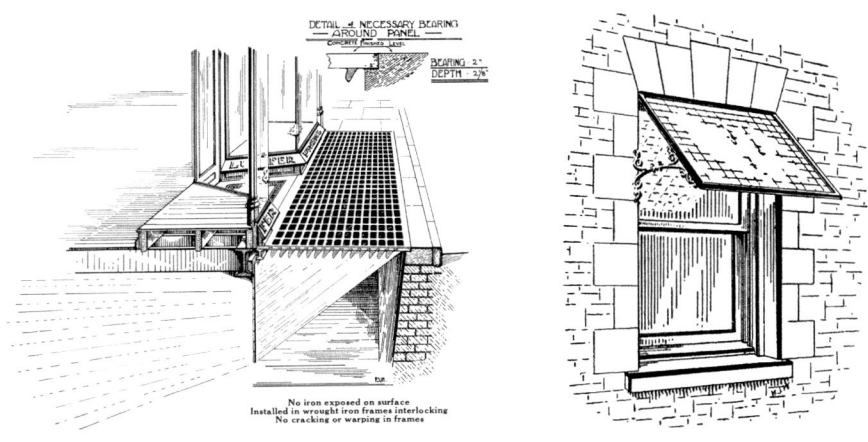

Figure 4.6
Above left: Luxfer Prisms used to refract natural daylight inside of a building are placed in 4" x 4" block grids above the plate glass windows of a typical c. 1900s shop front. Center: A Luxfer Prism diagram showing how daylight is re-directed indoors to the darker areas of both above ground and basement spaces. Source: Sweets Catalogue of Building Construction 1910.

Figure 4.7
At right: Luxfer Prism awning shown in a light shaft to deflect light indoors. Altogether, Luxfer Prism passive daylighting technology developed at the end of the 19th century was subsequently abandoned in favor of electric lighting, dropped ceilings with HVAC and changing architectural tastes. Source: The Luxfer Prism Co., Ltd. Of Canada, ca. 1910.

So popular was prism glass and glass block that more than 30 pages were devoted to the technology in the 1910 *Sweet's Catalogue of Building Construction*.[10] Architecture projects across the United States and Europe used variations of this innovation for natural illumination.[11] For example, New York City's original Pennsylvania Railroad Station, completed in 1910, used glass blocks to bring light from vaulted ceiling skylights down through the main concourse floors onto the railroad platforms below ground. After the station was demolished and modernized in the 1960s, many of these glass blocks still remain, but they are covered over from above.[12] Remarkably, much of the original Pennsylvania Station in New York was naturally lit during the daytime.

Figure 4.8
Original floor glass blocks (now covered over) used to direct natural light from the original 1910 Pennsylvania Station's vaulted skylights to the subterranean train platforms below. Electric lights were later expanded and now used on the platforms all day long. In 2021, the Moynihan Train Hall replaced the 1963 station for Long Distance Trains. Source: Shorpy.

Well-known designers of the time helped bring new patterns and aesthetic options to the appearance of prism glass and glass block, including Frank Lloyd Wright. Other more traditional architects derided them for not fitting into Beaux Arts and other neo-classical architecture fashions of the time. Towards the mid 20th century, electric fluorescent lighting allowed office buildings of all kinds to be thoroughly illuminated.[13] While fluorescent light proved more energy efficient than incandescent light, the cooler light quality is often unfavorable.[14] Because they contain mercury, broken fluorescent tubes are also categorized as hazardous waste.[15] Nevertheless, prism glass and related products, including glass block, had all but vanished from construction specification; transoms were removed, covered or painted over and dropped ceilings were installed in building interiors to enclose electric conduit and HVAC systems. In essence, these passive technologies were abandoned and are now mostly forgotten. Today, it is not uncommon to see surviving prism glass panels on façades from the early 20th century, but they are seldom used for their original purpose. In general, transoms, which helped extend the reach of natural light indoors, are not nearly as common as they once were, but they are still specified. Some new window films, offered by companies like 3M, offer refractive daylighting benefits similar to prism glass, but have had limited success.[16] Additional companies are developing glass blocks (and glass) with integrated photovoltaics to provide daylight while collecting solar energy.

Vertical Indoor Lighting Techniques: Light shafts

As obvious and ubiquitous as they may be, light shafts have played a long and significant role in the history of architecture, notably with the Roman Atrium.[17,18] Serving multiple functions, including internal daylighting, aeration and cooling, by the late 19th century, interior court spaces and staircases with central shafts served the dual purpose of both drawing warm air upward but also allowing fresh air and light downward into an internal space. Notable landmarks like the Temple Court Building in New York and the Rookery in Chicago have their internal plans designed around these important hollow features. As common as these interior architectural techniques were, they are less common in more recent buildings today. After all, they do encroach on usable, rentable space which can be illuminated with electric light and aerated with HVAC systems. However, when the Rookery was completed in 1888, Architecture critic Henry Van Brunt declared, "There is nothing bolder, more original, or more inspiring in modern civic architecture than its glass-covered court."[19] In this case, the Rookery's interior court pushed boundaries: it was clad in white brick to further reflect interior light throughout commercial spaces in the building. Such a color was principally chosen for utility, as the outside of the building was made of dark brown, rough-cut sand stone, which was architecturally fashionable in the 1880s and early 1890s. Frank Lloyd Wright, who understood the advantages of central light courts having worked in Chicago during the time when the Rookery was built, was later retained to renovate the interior in the early 1900s. Wright also introduced this type of interior court to many of his significant buildings, including the Larkin Building (Buffalo, NY, 1905, demolished) and much later, the Solomon R. Guggenheim Museum (New York, 1959).

Figure 4.9
Interior courtyard of the Rookery designed in 1888 by Burnham and Root conveys the illuminating effect of the light court's lightly colored brick beyond the glass. This technique was used to amplify natural light into the ground floor and suites above before the widespread availability of electric lighting which was added later. Source: Velvet (CC BY-SA 3.0).

Parallel to the use of light courts in commercial buildings, the light shaft played a similarly important role in residential urban architecture in the Western world. Echoing Covid-19-era concerns about the risks of stagnant air, health officials in the 19th century believed diseases spread much in the same way.[20] Daylight, too, was also believed to have health benefits. In cities like New York, single family rowhouses, lacking windows on the sides, began to incorporate interior light shafts to bring daylight and air into the center spaces of the house. In crowded low-income apartment houses, concern about rooms without windows led to the enactment of "tenement laws," or building codes that sought to ensure all inhabitants had access to light and air.[21] The first Tenement House Act passed in 1867 required windows in every room. This effort mostly failed. The second "Old Law" of 1879 was more successful by requiring functioning interior air shafts cut out from the plan of the building. The result was that every interior room could have a window and thereby fresh air. In response to the passage of the second law, trade magazine *Plumbing and Sanitation Engineer* sponsored a competition to design a dwelling scheme translating these requirements into reality. New York Architect James E. Ware penned the winning design which came to be known as the "dumbbell tenement" because of the resulting characteristic dumbbell-shaped plan. Unfortunately, many of the good intentions of the 1879 law and dumbbell design also backfired. The inaccessible ground floor of light shafts became receptacles for trash and human waste, creating additional (and unforeseen) health concerns. Fires were also able to spread quickly from one apartment to the next through air shafts.[22] In 1901, when the tenement laws were revised again, many of these health issues were resolved by widening the air shafts into courtyards and by allowing entry into the ground floor such that they could be cleaned and maintained. Afterward, courtyards and their variants continued to be used in different embodiments through the 20th century and beyond to bring natural light and air indoors. Early planned 20th-century communities, like Jackson Heights, New York, which was envisioned for the middle class, have apartment houses designed around central landscaped courtyards. Advertised as "garden apartments," these structures drew inspiration from British urban planner Ebenezer Howard.[23] About the same time, Swiss-French architect Le Corbusier's proposed Ville Radieuse for Paris, a design with a different configuration, but also emphasizing open space, greenery, light and air.[24] Today, large open courtyard spaces are valued, but are less common as mechanical ventilation and cooling have offered spatial alternatives.

Horizontal Indoor Daylighting: Transoms and Obscured Glass Walls

While large windows, skylights, light shafts and prism glass were able to direct natural light indoors, additional techniques have been employed to transmit light further inside into private, subdivided office spaces. Offices and factories often made wide use of translucent interior partitions, doors and transoms so that light could pass from outdoor windows into inner spaces. After the arrival of fluorescent light in the 1930s, electric indoor illumination became especially inexpensive and abundant. It could also be used in place of partitions and transoms for security reasons: burglars could easily break into rooms by simply breaking the glass on an obscured glass door or wall. Transoms, too, have also been abandoned, at least in part, for reasons of security. During the 20th century, they too became an easy access point for thieves. This disadvantage could be partially solved if the proportions of transoms were revised to further dissuade would-be thieves from passing through. Nevertheless, the value they bring by admitting daylight (and thereby saving electricity) has been partially overlooked. In contrast to more traditional office spaces with enclosed offices, open-plan offices, developed in stages in the 20th century, enable abundant interior transmission of light.[25] The disadvantage of the open-plan office is that it undermines privacy, which is sometimes needed for concentration and productivity. How could privacy be reintroduced today without interfering with horizontal daylighting? Some manufacturers that specialize in glass making have been reintroducing obscured glass partitions, some that can be moved around a space to promote needed privacy. Skyline Design, for example, hired pre-eminent French designers Ronan and Erwan Bouroullec to bring a fresh perspective to these interior glass partition elements. Together they developed an elegant solution through the *Oblique* and *Chevron* collection, which provides colorful, patterned natural light without compromising privacy.

Figure 4.10
Oblique and Chevron textured glass partitions designed in 2019 by studio Bouroullec for Skyline Design. Similar in effect to the textured glass used in obscured glass doors, walls and transoms in the early 20th century, they help harness natural daylight in open office spaces without compromising a sense of privacy. Source: studio Bouroullec.

Emerging Methods to Harness Natural Light: Sun Tunnels to Fiberoptics

As mentioned, in addition to promoting work productivity, exposure to natural light has been connected with mental wellbeing and physical health.[26,27] Light shelfs, sun tunnels, solar tubes, heliostats and other new (and reinvented)[28] passive light refracting and channeling technologies have been explored by numerous design firms, notably James

Carpenter Design Associates (JCDA), who specialize in large scale daylighting interventions in buildings. One notable example is their contemporary take on a light court designed for the New York Metropolitan Transit Authority's Fulton Street Transit Center. In this case, light is reflected from the outside down the light court into a public space below using tessellated reflecting panels.

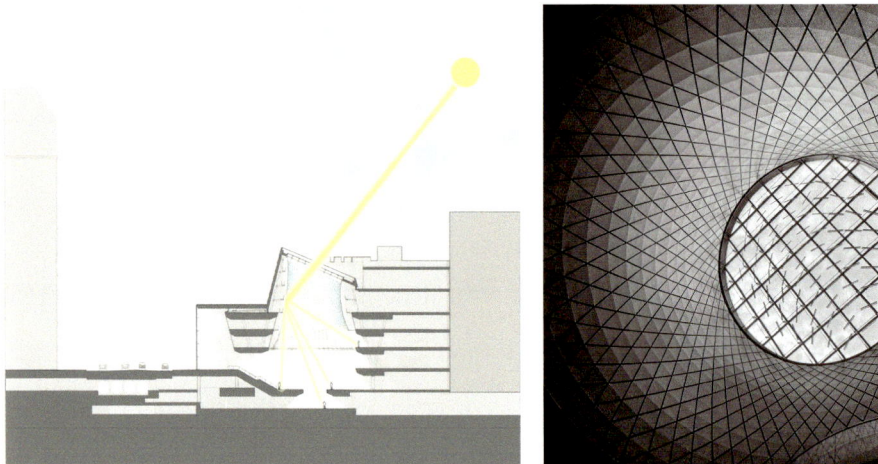

Figure 4.11
Above left: A schematic section view of the Fulton Street Transit Center 2004–2014, New York (Courtesy of JCDA). Source: James Carpenter Design Associates/ARUP/Grimshaw Architects/Carpenter Norris Consulting.

Figure 4.12
Above right: The Fulton Street Transit Center's central light reflecting court. Source: Brook Kennedy.

Fiberoptic Filament

Fiberoptic filament, used in telecommunication industries, has also found potential uses in daylighting buildings just as Luxfer Prisms did over a hundred years ago. In one recent example, James Ramsey and partners from architecture office studio RAAD developed a concept for an entirely daylit underground park in New York City called the "Lowline." Playing off the name of New York City's celebrated High Line Park, the "Lowline" proposed a subterranean landscaped park housed in a former underground electric streetcar terminal. After gaining considerable technical knowledge working at the National Aeronautics and Space Administration (NASA), Ramsey devised a heliotropic (sun-tracking) light collector able to channel sunlight underground to sustain a lush, green space. The project received extensive media attention and initial financial support, especially after a prototype installation (the Lowline Lab) was opened in late 2012. Studio RAAD continues to develop the technology as of the writing of this book but the future of the Lowline park itself remains uncertain. The Lowline Lab closed in 2017. Such daylighting technologies merit further investment for other applications. They could be used far beyond the Lowline. Just as glass block was used extensively in early New York City subway stations more than one hundred years ago, the Lowline Lab's daylighting technology (amongst other variants) would be a welcome addition to metro systems and underground spaces around the world today.

Figure 4.13
Above left: The fiberoptic Lowline concept prototype in 2012 demonstrating similar daylighting principles used in buildings before widespread electrification more than 100 years earlier. Source: Bit Boy (CC BY 2.0).

Figure 4.14
Above right: An 1881 patent for William Wheeler's light channeling invention. Source: USPTO.

Reflecting Daylight with Heliostats

Perhaps ever since the Siege of Syracuse (c. 214–212BC), when Archimedes allegedly set enemy ships ablaze using sunlight deflected from mirrors, humans have been trying to use mirrors and highly reflective material to direct light, including for supplementing illumination in interior spaces. Heliostats are a more contemporary invention which use mirrors to redirect light for collecting solar energy but also, in some cases, to illuminate shadowed spaces in towns and cities. In Rjukan, Norway, where the town center is cast in perpetual shadow during winter months, heliostats mounted high on a ridge above the town reflect sunlight into the central square.[29] In New York City, Carpenter Norris Consulting placed heliostats on a residential tower in Battery Park City to illuminate Tear Drop Park below, which otherwise receives little direct sunlight during the day.[30] Using robotic and adaptive control, heliostats could be more widely used and designed to harness daylight indoors and out to supplement or perhaps even replace electric lighting during the day in some conditions. Nearly two hundred years ago, in the 1850s, inventor Paul Chappuis patented a mirrored light reflector design that could be mounted on outside windows and adjusted manually to reflect natural light indoors.[31] Today, similar devices could be reintroduced to adjust their angle, following the sun's path. Whether using simple mirrors like Chappuis' prism glass, light channeling technology or interior translucent partitions, numerous further opportunities abound to leverage daylighting in tomorrow's built environment, for energy savings, health and human enjoyment.

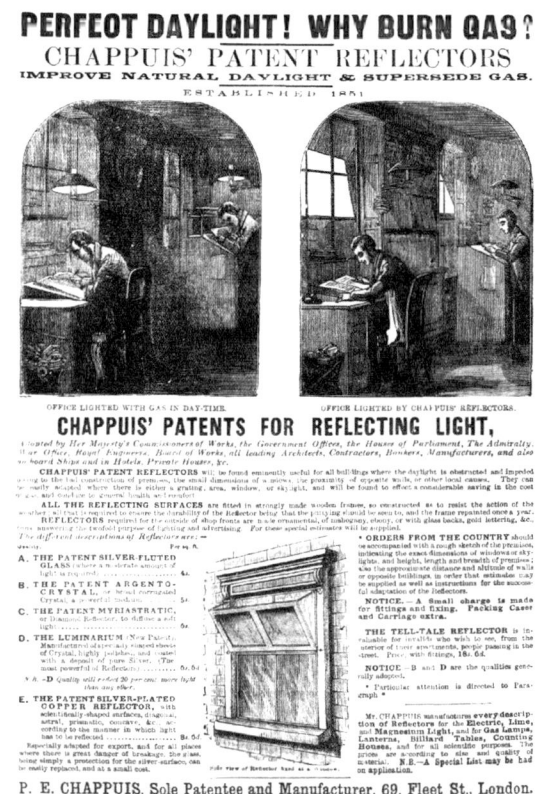

Figure 4.15
An advertisement for a Chappuis Patent Reflector, used for redirecting light indoors for dark interior spaces. Notice the angled window-mounted mirror illustration below. Source: Kelley's Directory of Wiltshire 1880.

Notes

1 US Energy Information Administration (2022) 'How much electricity is used for lighting in the United States?' Available at: www.eia.gov/tools/faqs/faq.php?id=99&t=3 (Accessed: November 19, 2021).

2 Osibona, O., Solomon, B.D. and Fecht, D. (2021) 'Lighting in the home and health: A systematic review,' *Int. J. Environ. Res. Public Health.* 18(2), p609. Available at: www.ncbi.nlm.nih.gov/pmc/articles/PMC7828303/ (Accessed: August 9, 2022).

3 Chepesiuk, R. (2009) 'Missing the dark: Health effects of light pollution,' *Environ Health Perspect,* 117(1). Available at: www.ncbi.nlm.nih.gov/pmc/articles/PMC2627884/ (Accessed: August 9, 2022).

4 Tanizaki, J. (1977) *In praise of shadows.* New York: Leete's Island Books.

5 OpenLearn (2019) 'Lighting the industrial revolution,' *The Open University,* August 30. Available at: www.open.edu/openlearn/history-the-arts/history/history-science-technology-and-medicine/history-technology/lighting-the-industrial-revolution

6 Larger municipal gas lighting systems have been preserved in a few cities around the world including London, Prague, Berlin, Boston, Cincinnati and several smaller townships in New Jersey, including Glen Ridge, South Orange, Palmyra, Riverton and others. Riverside, Illinois, a community planned by Frederick Law Olmstead, retains its original gas street lights. Brooklyn New York's Park Slope Historic District maintains gas lamps on the properties of its elegant rowhouses.

7 National Park Service (2015) *The electric light system* [online]. Available at: www.nps.gov/edis/learn/kidsyouth/the-electric-light-system-phonograph-motion-pictures.htm (Accessed: August 20, 2022).

8 The New York Historical Society (2014) *When Edison lit up Manhattan*, September 4 [online]. Available at: http://behindthescenes.nyhistory.org/edison-lit-manhattan/ (Accessed: September 9, 2022).

9 Howe, B.R. (2023) 'New Skyscraper, Built to Be an Environmental Marvel, Is Already Dated,' The New York Times, February 14 [online]. Available at: https://www.nytimes.com/2023/02/14/climate/green-skyscraper-one-vanderbilt.html (Accessed: February 16, 2023).

10 Sweet's Catalogue of Building Construction (1910) [online]. Available at: https://babel.hathitrust.org/cgi/pt?id=uiuo.ark:/13960/t3710nk4j&view=1up&seq=702&skin=2021&q1=prism%20glass (Accessed: September 9, 2022).

11 Newman, D. (1995) 'The century's triumph in lighting: The Luxfer Prism companies and their contribution to early modern architecture,' *Journal of the Society of Architectural Historians*, 54(1), pp24–53.

12 Today, nearly 60 years after the original Pennsylvania Station was demolished, the Moynihan train station opened across the street with a large central skylight placed above the main concourse.

13 Bright, A.A. (1949) *The electric-lamp industry: Technological change and economic development from 1800 to 1947*. New York: Macmillan.

14 Hamilton, W.L. (2007) 'Incandescence, yes. Fluorescence, we'll see,' *The New York Times*, January 7 [online]. Available at: www.nytimes.com/2007/01/07/weekinreview/07hamilton.html (Accessed: August 9, 2022).

15 US Environmental Protection Agency (2021) *Recycling and disposal of CFLs and other bulbs that contain mercury.* Available at: www.epa.gov/cfl/recycling-and-disposal-cfls-and-other-bulbs-contain-mercury (Accessed: August 9, 2022).

16 Titan (n.d.) *3M Daylight Redirecting Film.* Available at: www.titanwindowfilms.com/3m-sun-control-solutions/3m-daylight-redirecting-film.html

17 Steemers, K. (2000) *Architecture, city, environment.* Cambridge: Earthscan Press.

18 Sharples, S. and Shea, A.D. (1999) 'Roof obstructions and daylight levels in atria: A model study under real skies, *Lighting Research and Technology*, 31(4), pp181–185. doi:10.1177/096032719903100408.

19 Cited in Conditt, C.W. (1964) *The Chicago School of Architecture.* Chicago: University of Chicago Press, p64.

20 Allen, J.G., Spengler, J. and Cedeño-Laurent, J. (2020) 'Want to buy schools time? Open the Windows,' *The Washington Post*, August, 27 [online]. Available at: www.washingtonpost.com/opinions/2020/08/27/want-buy-schools-time-open-windows/ (Accessed: August 9, 2022).

21 Plunz, R. (1990) *A history of housing in New York City.* New York: Columbia University Press.

22 DeForest, R.W. and Veiller, L. (eds.) (1903) *The tenement house problem: Including the report of the New York state tenement house commission, in two volumes.* New York: MacMillan.

23 Howard, E. (1898) *Garden cities of to-morrow.* London: Swan Sonnenschein & Co.

24 Merin, G. (2013) 'AD Classics: Ville Radieuse / Le Corbusier,' *archdaily.com.* Available at: www.archdaily.com/411878/ad-classics-ville-radieuse-le-corbusier/ (Accessed: August 9, 2022).

25 Musser, G. (2009) 'The origin of cubicles and the open-plan office: Wall-free office spaces did not quite work out the way their utopian inventors intended,' *Scientific American*, August 17 [online]. Available at: www.scientificamerican.com/article/the-origin-of-cubicles-an/ (Accessed: August 9, 2022).

26 Lockley, S.W. (2009) 'Circadian rhythms: Influence of light in humans,' in Larry, R.S. (ed.) *Encyclopedia of neuroscience.* Oxford: Academic Press, pp. 971–988.

27 Kontadakis, A., Tsangrassoulis, A., Lambros, D. and Zerefos, S. (2018) 'A review of light shelf designs for daylit environments,' *Sustainability*, 10(1), 71.

28 Dabija, A. (2017) 'Building with the sun. Passive solar daylighting systems in architecture,' in Visa, I. (ed.) *Nearly zero energy communities.* Springer. DOI: 10.1007/978-3-319-63215-5_6. This article references sun tunnels produced by Velux. Several other companies at the time of this book's writing produce light shelf systems, solar tubes and sun light channeling technology.

29 Taylor, A. (2013) 'Using giant mirrors to light up dark valleys.' *The Atlantic*, October 22 [online] Available at: www.theatlantic.com/photo/2013/10/using-giant-mirrors-to-light-up-dark-valleys/100613/ (Accessed: August 9, 2022).

30 Dumiak, M. (2007) 'Heliostats tap sunlight for lighting outdoor and indoor spaces,' *Architectural Record*, May 19 [online]. Available at: www.architecturalrecord.com/articles/6674-heliostats-tap-sun light-for-lighting-outdoor-and-indoor-spaces (Accessed: December 7, 2021).

31 Gritsiyenko, D. and Kasyanov V. (2017) 'Solar exposure condition improvement in urban area using light guide,' *MATEC Web of Conferences*, 106. Available at: www.matec-conferences.org/articles/ matecconf/abs/2017/20/matecconf_spbw2017_01015/matecconf_spbw2017_01015.html (Accessed: August 9, 2022).

Chapter 5

The Hidden Costs of Domestic Upkeep

This chapter explores methods of domestic cleaning and maintenance, especially with energy and resource intensive areas of cleaning activity: clothing driers, dishwashers and vacuum cleaners to offer a few examples. All of these industrialized appliances consume energy to make life more convenient and to keep pace with contemporary standards of hygiene. By reducing human effort involved in household chores, their intention is to liberate families (especially stay at home parents) to do other things, especially in service of social mobility. Meanwhile, these conveniences, like AC, come at an environmental cost, especially when used in isolation. Energy use in US homes, fueled in part by smaller electric devices and cleaning technologies, has contributed to increasing per-capita energy consumption, and thereby carbon emissions.[1] In this chapter, alternative and passive technology will be considered to offer new directions for energy and water-saving design, reduced consumer product waste and lower maintenance.

The Social Promise of Labor-Saving Cleaning Technology for Women

It is impossible to discuss the development of 20th-century home cleaning technology in the West without addressing the significant role it has played on the lives of women, who performed (and continue to perform) most of the cooking, laundry and house cleaning at home.[2] This remains true even as the number of women entering the workforce full-time has risen substantially since 1950.[3] In *Never Done: A History of American Housework*, Susan Strasser's social history of home cleaning demonstrates how the accessibility of new home cleaning technologies reduced the amount of physical labor, particularly in the lives of women and especially after the Second World War.[4] Other social histories have been less optimistic. Ruth Cowan's landmark socio-technical history, *More Work for Mother: The Ironies of Household Technology from the Open Hearth to the Microwave*, argues how many new industrialized home technologies (the spin dryer, range, vacuum cleaner, dishwasher, sewing machine, etc.) were not entirely successful in saving women time and effort. Instead, Cowan asserts that these aforementioned home technologies only *entrenched* time spent cleaning in the home by creating ever higher standards of domestic hygiene.[5] In effect, cleaning standards which were once tolerated in the early 20th century were no longer excusable in the electric washer-dryer era that followed. To be sure, by evaluating the comparative energy efficiencies of past domestic technologies, this book is by no means suggesting a return to the gender inequities connected with them. In contrast,

DOI: 10.4324/9780367814304-6

these alternative and traditional technologies are presented as energy saving opportunities for home owners, and encourages designers to find new ways to integrate their use in today's homes.

Hang Drying Clothes vs. the Electric or Gas Tumble Dryer

In an average American household, electric and natural gas-powered tumble driers consume as much as 5% of a home's monthly energy budget.[6] Before these machines were widely available in the 1950s, clothes were often hung dry in back yards, basements, across narrow alleys, in bathrooms, kitchens, basements or wherever clothes could be practically hung. Of course, hang drying is still practiced widely today, in some countries more than others, where energy is more costly and where these appliances and the energy that fuels them is unavailable. Unsurprisingly, the US is a leader in energy consumption in this sector even though air drying is still encouraged at the federal level.[7] Beyond energy savings, further benefits can be cited for air drying: driers can damage delicate fabrics over time. As a result, some garments can last longer when air-dried. Further, hang drying also prevents natural fabrics from wrinkling and some enjoy the laundered aroma they produce. At the same time, new drier technology is improving. Heat pump driers, for example, common in Europe since the 1990s, can use up to 50% less energy than tumble driers in the US. As this technology finally makes its way across the Atlantic Ocean, households can still reduce energy use further by hang drying some garments as well.[8,9,10]

Heated Hang Driers

Much has been written about the benefits and drawbacks of clotheslines in cities and in suburbs.[11] By the late 19th century, hang drying technology developed further to work within the basements of apartment buildings in denser urban areas. Companies including Chicago Dryer and Judson, for example, developed energy and space-efficient hang-drying racks that used various configurations including existing building heating to accelerate drying speeds.[12] Some models used independent stoves that would also boil water for washing. Later models employed electric heating. Tenants of apartment buildings could hang their clothes privately in these thin basement racks, but as gas and electric tumble driers became available many of these hang-drying units were replaced with tumble driers.

Back Yard Hang Driers

One former design staple of the suburban home is the rotary, outdoor fold-out clothes drier. Permutations of this commonplace device could be found in American, European, Australian and Canadian yards especially after the Second World War. In Australia, the *Hills Hoist* evolved beyond a mere utilitarian device. In addition to serving its staple function to dry laundry, the spinning cantilevering frame adapted into a cultural phenomenon, including use for yard and even drinking games.[13] As Cowan and Strasser traced in their social histories, the electric and gas tumble drier slowly began to overtake the use of different kinds of clothes drying methods. Moreover, with the rise of "no iron" fabrics, hang drying continues to be a niche practice in many industrialized countries. In the late 20th century, hang drying began to be regarded as anachronistic and uncivilized in the US and Canada—an expression of poverty rather than bourgeois aspiration. Suburban home owners' associations began to ban hang driers in the 1970s, claiming their unsightliness lowered real estate property values.[14] Over time, more than dozens of American states and Canadian provinces had clothesline bans in place. As of 2020, twenty states have now overruled these local regulations with so-called "right-to-dry" laws. As a result, hang drying has made a minor come back. But perhaps the originators of the bans had a point. The design of these hang-drying devices is often uninspired—they have remained mostly unchanged since they were introduced over a century ago. To encourage their acceptance and use today, there is certainly an opportunity to reinvigorate the design of these rotary deployable hang driers, using contemporary fabrication and durable new materials as German manufacturer Juwel has done for indoor racks, to otherwise advance their public image beyond cheap utilitarian accessory. Could they not be a durable suburban essential? Like with many tired consumer product categories, design could play a pivotal role reinventing these passive, energy-conserving devices.

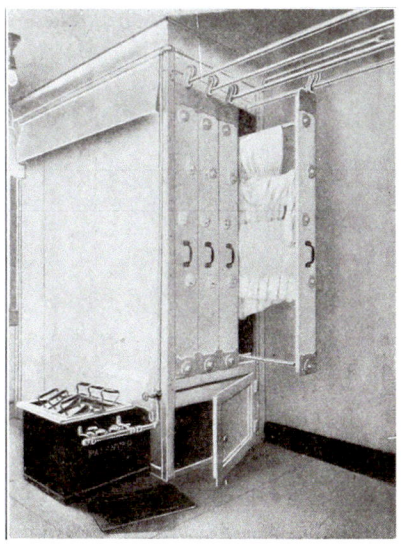

Figure 5.1
Above left: A 1910 Chicago heated hang-drying compartment shown in an apartment building basement. These were commonly used in the early 20th century before the arrival of the tumble drier and likely used less energy. Source: Sweets Catalogue 1910.

Figure 5.2
Above right: A contemporary indoor fold-up laundry hang dryer by German manufacturer Juwel. Hang drying has long been an alternative method of drying clothes that uses far less energy than tumble driers. Source: Juwel.

Hang-Drying Kitchen Cloths vs. Paper Towels
Each year more than 50,000 trees are cut down to produce the millions of tons of paper towels that are consumed and discarded in the US alone.[15] Invented by Clarence and Irwin Scott in 1879, the paper towel took decades to evolve into the ubiquitous roll used in homes today. Before their invention, reusable, cloth kitchen towels were dominant. In many Western countries

they continue to be used, even though the disposable paper towel continues to gain ground. Past devices and designs demonstrate clever ways of allowing laundered or wet cloth kitchen towels to dry. Patents from the USPTO reveal numerous earlier inventions to hang-dry kitchen towels overhead, either after they have been used to clean or after they have been washed in the laundry. Often handles on refrigerators, ovens and cabinets double as places to dry cloth towels, some not very effectively. How could kitchens today be consciously designed to accommodate reusable kitchen towels thereby making them more convenient and dominantly used again? Contemporary product designers like Katrin Greiling address this question through her project *Extra Hands*, a beautiful, humble tool holder (and room divider) designed to honor basic domestic tools and the cleaning chores associated with using them, including kitchens towels. In Greiling's words, *Extra Hands* reimagines "the theme of chores, conventionally understood as tedious, burdensome work to be avoided." She goes on by proposing:

> a renewed sense of engagement between humans, objects and their environment in the pursuit of a more virtuous way of living through ritualizing the mundane, with the potential for clearing one's mind, connecting with one's body and confronting the indifference of the natural world.[16]

Designed for the Furnishing Utopia 3.0 *Hands to Work* exhibit during New York Design Week in 2018, *Extra Hands*' sensible design and durable materials help restore nobility not only to the design of the rack and cloth towel but also to the act of cleaning.

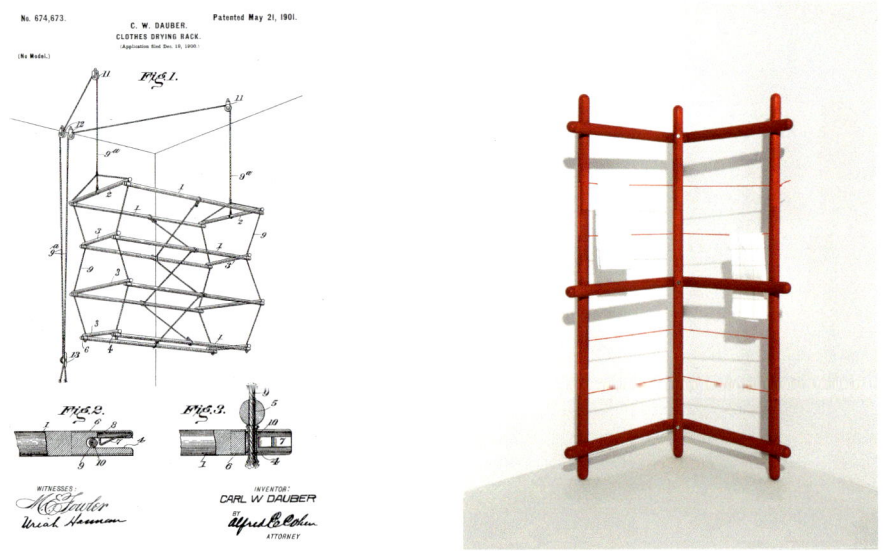

Figure 5.3
Above left: A once common and patented lowering drying rack for laundry and the kitchen, patented in the early 20th century and used in the US, England and elsewhere. Source: USPTO.

Figure 5.4
Above right: A contemporary space-saving kitchen drying rack designed by German Industrial designer Katrin Greiling for Furnishing Utopia 3.0 – Hands to Work exhibition at New York Design week in 2018. Source: Katrin Greiling.

Laundry Mangles

Long before electric washers and driers were introduced in the 1930s, additional mechanical tools to help wash and dry laundry were used. One particularly cumbersome device was the washboard, a corrugated piece of sheet metal over which wet soapy garments would be rubbed to release dirt, debris and stains. These were used inside wood or metal tubs full of water and required considerable repetitive effort to be effective. Another device worth consideration was the laundry mangle, an invention dating to the late 16th century. Instead of using electricity to produce a centrifugal force to extract water from laundry (often called the "spin" cycle since their introduction in 1937), manual mangles used mechanical advantage to squeeze water out of fabrics between two rollers.[17] Some early washing machines even had integrated powered mangles that would press water out of clothing after coming out of the wash cycle. This process, combined with hang drying, used far less electricity than an electric dryer. Electric mangles gained notoriety for pinching fingers that unwittingly or accidentally were pulled between the rollers. Some have even wrongly speculated that the origin of the term "mangle" originated from hand injuries from these machines ("mangled fingers"). Ultimately, the electric varieties fell quickly out of favor once convenient electric driers entered the marketplace. Today, powered mangles continue to be used for pressing and ironing laundry by dry cleaners, laundering facilities and in some residential environments especially outside the US.

Figure 5.5
A Miele electric washer with a powered mangle. Mangles are no longer commonly used for drying but remain in service for pressing and ironing large flat textiles like curtains and bed sheets. Source: Andreas Praefcke.

On the other hand, manual, geared mangles have been used for centuries without electricity and continue to be used in niche applications—they have the advantage of minimizing finger pinching but require some physical exertion—seemingly the anathema of modern civilization. Like Katrin Greiling's *Extra Hands*, perhaps the mangle could be redesigned to better integrate into home cleaning tasks to avoid extra drying loads and energy use. Heavier absorbent fabrics, like sweatshirts, could benefit from being wrung out with a mangle prior to entering an electric drier to avoid hours-long drying cycles.

Cooking and Food Preservation

Cooking has long overlapped with home heating. Once, fireplaces were also used for cooking, until eventually stoves became individual appliances made of cast iron, brick and other materials. Wood, biomass, coal and other materials were gathered and used for fuel. Later, these fuels were replaced with natural gas and electricity. As the cooking stove developed into the appliance known today, so did diets and the timing of daily meals. Often winter meals like stews and soups are cooked over time and kept warm using low heat, to avoid reducing the dish excessively or drying it out. Meals, whether slow cooked or otherwise, require sustained energy use; as researcher Kris De Decker astutely observes, "The (gas stove) cooking process is similar to heating an uninsulated building with all the doors and windows open."[18] However, this was not always so. In the past, several cooking techniques that are seldom used today could easily be reintroduced for energy savings. Those that are still used are mostly unknown but could be reintroduced today using contemporary materials and updated design.

Fireless Cookers

So-called *fireless cookers* are probably inappropriately named since they give the impression that they use no energy at all. On the contrary, they continue to cook food for hours after being inititally heated, generally over some fire source. However, once heated, it is the insulation of the food pot that keeps it warm inside and continues to cook the contents for hours without additional heat. Evidence of these kinds of cooking devices have been discovered in civilizations around the world including Norway and in the settlements of numerous indigenous American cultures including the Havasupai of Cataract Canyon, Arizona, the Pomos in California and the Eskimo.[19] There are several examples that use different types of insulation or heat containment, but in the 19th century they developed into industrial consumer products. During the Paris Exposition of 1867, a Norwegian fireless cooker was demonstrated. Later the invention was described in the science journal *Scientific American*. Called a "Self-Acting Norwegian Cooking Apparatus" in the September 11, 1869 issue, the article claimed that,

> the heated vessels containing the food will retain a high temperature for several hours, so that a dinner put in the apparatus at eight o'clock in the morning will be quite hot and ready by five in the afternoon and would keep hot until eleven or twelve at night.[20]

Here again, it is the insulation not a continuous heat source which is doing the cooking, which Kris De Decker further describes as the "Passive House concept applied to cooking."[21]

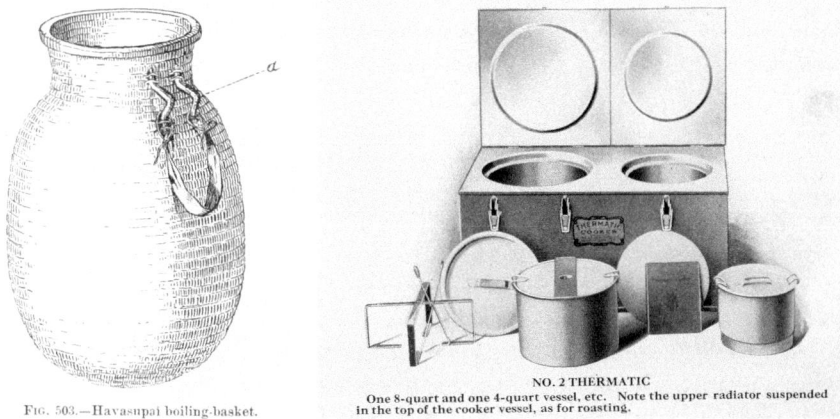

FIG. 503.—Havasupai boiling-basket.

NO. 2 THERMATIC
One 8-quart and one 4-quart vessel, etc. Note the upper radiator suspended in the top of the cooker vessel, as for roasting.

Figure 5.6
Above left: A Havasupai boiling basket, used with heated stones, insulating hides and sometimes clay pot enclosures. Source: Smithsonian Libraries.

Figure 5.7
Above right: A Thermatic brand fireless cooker resembling the Norwegian cookers of the mid-19th century. Modern insulating materials could further improve the energy efficiency of these already low-energy cooking appliances. Source: Sheer, H. M., company, Quincy, Ill. [from old catalog].

Chambers Range
Model 3742-W

This range is furnished with either right or left-hand oven finished either in black and white as shown on page 12, or all white as illustrated. The oven dimensions of this range are 17 x 17½ x 12. The size of the cooking top, 29 x 26; cooking top height, 31 inches, and gross dimensions of range are: Width, 54 inches; depth, 31 inches; height, 59 inches.

3742-WR (Illustrated). Two Thermodomes, fireless oven and broiler, four top burners and simmer. Oven on right, finished in all white.
Code . . . . . . Eclair
3742-WL Same as above, finished all white, oven on left.
Code . . . . . Eclipse
3742-BR Same as above with black and white finish shown on page 12. Code . Eccentric
3742-BL Same style and finish as 3742-BR except with oven unit on left. Code . . Echo
Shipping weight of all above models, 590 pounds.

§ 8 ½

Figure 5.8
Chambers Range with two "Thermodomes." Later stove models such as the General Electric "Deep Well" cooker integrated insulating features into a more conventional cook top. Source: Chambers Manufacturing Company. (Undated, Late 1920s). The Chambers Fireless Gas Range.

By the 1910s, these devices began to be integrated into conventional kitchen ovens and ranges. John E. Chambers, for example, developed the "Fireless Gas Range" and his eponymous Chambers Stove Company in 1912. Chamber's kitchen stove featured stove-top covers which could be lowered over pots of slow cooking or cooked food to insulate and save energy.[22] Other models featured insulating interior compartments. By the 1930s, Hotpoint and General Electric offered insulated stove-top wells to keep the contents of a cooking pot warm without needing additional energy. Today, manufacturers such as Kuhn Rikon and Thermos continue to sell descendants of these earlier fireless cookers to niche markets using more advanced contemporary materials. Sometimes fireless cooking is used in the developing world, for camping or off-grid scenarios. Slow cookers and crockpots, on the other hand, have remained popular, for reasons of convenience. Originating in the 1940s they are less effective at insulating and rely on sustained electric heat to cook. Still, one can imagine insulated cooking chambers reemerging in modern cooking appliances today, perhaps even with induction cooktops, combining new materials and manufacturing processes for convenience, safety and improved energy efficiency.

Food Preservation and Refrigeration

Civilizations have long sought ways to preserve food. The ability to store food rather than having to hunt or harvest it daily has allowed humankind to stay in one place and form prosperous communities. Interestingly, cultures from around the world developed many of the same food preservation techniques depending on their local climate.[23] For example, climates in colder weather regions learned to cool and freeze their food whereas cultures in warmer climates often learned to dry their foods. Additional techniques such as curing (with salt and sugar), pickling and fermenting have helped preserve food for subsequent consumption. Later in the 18th century, canning was developed as an additional effective method. In contrast, mechanical refrigeration has become the dominant method of preserving food today in industrial economies, differing from past methods in its substantial energy consumption. Technologically similar to AC, refrigeration makes life more convenient. Produce, meats, fish, dairy products and other perishables can travel long distances without spoiling. While this chapter is not focused on the many resourceful traditional techniques for preserving food, we will briefly touch on some passive traditional methods built into the home and kitchen environments that have helped perishable food last longer.

Cellars and Cold Pantries

Cellars were once used more widely to store foods, especially fruits and vegetables. In their placement below the surface of the ground, the temperature is usually cooler than outside or the upper floors of a house. Root cellars are often separate buildings used for the same purpose, specifically for the preservation of root vegetables which can last weeks or months inside these spaces. Other passive cooling structures for preserving food include cold pantries, larders and spring houses which can use thick stone insulating walls to maintain cooler temperatures. Spring houses are further cooled by running water emerging indoors from natural springs. Today, the traditional cellar and related structures have slowly evolved away from food storage into recreational rooms, additional bedrooms, storage areas and other non-food preservation related purposes. In the kitchen, evaporative cooling and other techniques for preserving food without energy have long been used. Designers Ellie Perry and Jihyun David have re-envisioned different traditional food preservation technologies, with contemporary materials and aesthetics. Perry's Thermacooler is based on *Zeer Pots* and other Middle Eastern and African precedents.[24] David's is based on traditional Korean knowledge and techniques to preserve vegetables.

Figure 5.9
Top left: A pond used at Thomas Jefferson's Monticello for keeping fish alive rather than preserved using other means. Source: Brooklyn4083 (CC BY-SA 4.0).

Figure 5.10
Top right: An energy-free food preservation system proposal by industrial designer Jihyun David called *Save Food from the Fridge.* **The system borrows from traditional Korean practices and knowledge about food preservation, showing here vertical orientation. Source: Studio Jihyun David, www. savefoodfromthefridge.com.**

Figure 5.11
Bottom left: Designer Ellie Perry's 2022 *Thermacooler* **based on ancient precedents like Zeer Pots, uses evaporative cooling techniques to preserve fruit, vegetables and other perishables. Source: Ellie Perry.**

Figure 5.12
Bottom right: Designer Rochus Jacob's Thermodynamic cooler also uses evaporative cooling techniques to preserve fruit, vegetables and other perishables. Source: Rochus Jacob.

Dust Collection: Vacuuming, Brooms and Carpet Sweepers

The vacuum cleaner has become such an integral feature of modern home and office cleaning that it is hard to imagine how this task was performed before electricity. By offering a convenient means of collecting floor debris, dust and dirt, vacuum cleaners now come in a variety of configurations: canister, uprights, handheld and more. For those with pets and allergies, they also collect hair, pollen, dander and "dust bunnies." In doing so, like many modern appliances, they consume energy. Always on, whether collecting much debris or not, an average vacuum cleaner can use more than 1000 watts (or the equivalent of 10 incandescent light bulbs while operating). In one household, this might not seem significant, but when multiplied across the industrialized world, cumulative vacuuming amounts to an enormous amount of energy consumption. Now, this book would certainly not suggest abandoning the practice of vacuuming and, indeed, vacuum cleaners are not used for long periods of time. Instead, they spend much of their time in the closet. However, when on, they are frequently running at a fraction of their capacity picking up tiny scraps spread out over a large area. Sometimes, people use them in combination with other cleaning tools such as brooms and dust mops, like Proctor and Gamble's electrostatic Swiffer, which can collect some debris on hard floor surfaces. Similar reusable, electrostatic "microfiber" dust mops have been available for some time as well. Neither consume energy.

Nevertheless, Swiffer and dust mops do not clean rugs or carpets successfully, nor are they suited to picking up larger crumbs or loose debris. (Swiffer's now-abandoned "Carpet Flick" was a noble attempt to produce a manual debris collector but it never caught on and, like the Swiffer, required using a disposable pad.) How else then can debris collection be achieved without using electricity and without a disposable pad? For thousands of years human beings have been removing dirt and debris from the inside of inhabited spaces without electricity, perhaps not as successfully. Before the arrival of the vacuum cleaner, many other tools were invented, some that persist today. The most obvious example is the broom and dustpan, a workhorse that can be used on any surface. Other tasks, such as rug beating, were commonplace and cumbersome: rugs are removed from a space, hung up vertically and beaten to loosen dirt and debris. Less known is the manual carpet sweeper, which we will discuss further on.

The Ephemeral Value of Brooms

Brooms and their bristle design have evolved over thousands of years and remain in use today alongside vacuums and other tools, even those controlled robotically. Bristles, unlike suction power alone, serve to loosen debris from carpets, floors and move them directionally; vacuum suction, on the other hand, employs fans or bellows to lift particles through brute force. Some evidence suggests that brooms have been used since the 2nd or 3rd Millenia BC in China. Scattered evidence suggests there has been limited development of the tool and its material components for hundreds of years. By the turn of the 19th century, there was a boom in broom making and development. In 1797, American farmer Levi Dickenson developed a durable and effective broom made from Sorghum fibers. By the 1850s, Dickenson's broom had become so popular and widespread that the bristle grain was often referred to as "broom corn." Additional influential modifications to the modern broom were made by automating the means by which the bristles were sewn and secured together and through methods of cutting the broom tips evenly. Modern designs continue to make brooms a part of the cleaning arsenal at home and outdoors, despite the development of vacuum technology. Brooms were also used in conjunction with other infrastructural aids built into homes, like fireplace ash chutes which would carry fireplace ash into basement receptacles for easier collection. Similar tools were developed for street cleaning until mechanized tools took their place.

"THE STEPHENSON" STREET SWEEPINGS RECEIVER.
FIFTH AVENUE AND BROADWAY, NEW YORK, N. Y.

Figure 5.13
Above left: A traditionally handcrafted Sorghum Broom with natural indigo dye designed by Karen Lane. Source: Karen Lane.

Figure 5.14
Above center: Traditional Filipino hand-made brooms. Source: Photo by CEphoto, Uwe Aranas (CC BY-SA 3.0).

Figure 5.15
Above right: Fireplace chutes are often used to help clean up dust by offering a trap door into which dust can be swept and then collected in the basement without needing a vacuum cleaner. Paired with a broom, "dust chutes" could be designed to collect household dust in the same way without electric energy use. In the first decade of the 20th century, manufacturer C.H. Stephenson designed an inset bin into which debris and dust could be swept. The bin could be lifted out of the ground when full for easier disposal.[25] Source: C.H. Stephenson.

The Manual Carpet Sweeper

Concurrently with the 19th-century development of brooms, horse-pulled mechanical sweepers were introduced in London to clean city streets that were often strewn with horse manure. Here, a rotational broom apparatus would turn to collect the manure and other debris. Not long afterward, a hand-powered variant of this design was patented in the UK. By pushing the device forward on a floor surface, the wheels would turn the rotational brush to sweep up debris indoors. Later, in the US, a similar design was developed by Boston inventor Hiram Herrick who submitted a patent for a "carpet sweeper" in 1858.[26] Herricks's sweeper was later evolved by Melville Bissell of Grand Rapids, Michigan who likewise patented an improvement to the carpet sweeper typology.[27] Like Herrick, one of Bissell's main focuses was to overcome the problem of debris and threads (and presumably hair) clogging the motion of the wheels. Patented carpet sweeper designs of 1876, 1881 and 1883 remain mostly unchanged today, sold in department stores and online retailers alike.

The Rise of the Vacuum Cleaner

Vacuum cleaning was also first envisioned before electric power. Beginning around 1860, inventors Daniel Hess of West Union, Iowa began producing suctioning machines powered by manual bellows. By the end of the 19th century, large steam powered vacuum machines were developed in the UK and the US which offered a vacuuming service to households. These machines were too big and produced too many fumes to be brought indoors.[28] Electric-powered vacuum cleaners that resemble today's domestic cannister and upright models were introduced in the first decade of the 20th century. In 1905, Walter Griffiths made the first "cannister" vacuum with multiple vacuum head attachments and a flexible hose. In 1907, janitor James Spangler of Canton, Ohio developed an electric-powered vacuum cleaner resembling upright models still familiar today. Cash poor, Spangler sold his patent to William Hoover in 1908, a leather goods manufacturer, who refined the appearance of the model and introduced it to the public just as houses were becoming electrified. Later in 1920, the company was renamed the Hoover company. Today, "hoovering" is now synonymous with the act of vacuuming in the UK and other countries. Fisker and Nielsen in Denmark and Electrolux in Sweden developed similar models at the same time. At first, "Hoovers" were a luxury purchase at $60 but as homes and offices increasingly gained access to electricity their use expanded. Huge efforts to electrify homes continued through the depression in the 1930s such that by the post-Second World War period most Americans could afford one. Nevertheless, as electric vacuum cleaners were adopted, homes continued (and continue) to rely on brooms with dustpans and other hand-powered devices. That balance has slowly eroded; mechanical carpet sweepers, invented in the 19th century, offer a plausible alternative and they continue to be used today, but are increasingly overshadowed by a newer generation of vacuuming innovation, notably Dyson's Cyclonic Technology and robotic vacuum cleaners from Electrolux, Neato and Roomba.

Are Carpet Sweepers Still Useful and Relevant?

Unlike vacuums, mechanical carpet sweepers use no electricity, and they are still commercially available. Could they compete in some instances with vacuums? Could they work in tandem with vacuums to help save energy? To help answer this question, an informal product evaluation test was conducted with 12 undergraduate Industrial Design students at a US university. The goal was to use qualitative design research methods to evaluate whether a newly purchased carpet sweeper (the brand will remain unnamed) could perform credibly. Six-inch diameter piles of sawdust were placed on a floor carpet in a university classroom to conduct the test. Users (15 students aged 21 and 22) were asked to test the carpet sweeper and evaluate their performance. After trying the new device, the students found the non-electric device perfectly capable of cleaning up the spilled sawdust—much to their surprise, despite initial and unanimous skepticism. When asked how the device could be improved, there were three principal suggestions: 1) improving the ease of emptying the collected dust, 2) improving performance in corners (when the sweeper does not move it can't pick anything up) and 3) the devices often looked cheap and unattractive. It is true; unlike aspirational Dyson and Miele vacuums, the carpet sweeper appears dated and unconsidered in terms of aesthetics, materials and feel: outmoded color schemes, cheap construction and grinding plastic sounds from the wheels and gearing. The conclusion was, just like many other alternative and traditional tools, the carpet sweeper begs to be redesigned for a contemporary, design savvy, sustainability-committed audience. What if the device used durable and/or renewable materials such as wood, recycled or biodegradable plastics, aluminum and others while ensuring it was still relatively light and easy to wield? What if the design aspired to an aesthetic matching the contemporary design languages of influential home and technology companies? What if they were designed to be left out proudly like a piece of high-end furniture or an Italian espresso maker? Beyond utility, design has the power to *present* technology in an emotionally engaging way—Dyson's cyclonic technology and Roomba's autonomous control, for example, is presented futuristically like the GM cars of the 1964 Futurama exhibit: an exciting new development.[29] Other contemporary manufacturers, like MUJI of Japan, excel in presenting timeless examples of familiar tools and technologies in a compelling way. For Dyson, their success is likely just as much an outcome of the *presentation* of their technology, not just the act of dust collection which ultimately drives their popularity.

Perhaps one irony is that carpet sweeping technology has already been partially adapted to one of the newest vacuum technologies—those with robotic control. In fact, some early robotic models, like the Trilobyte by Electrolux, have sweeping mechanism surprisingly reminiscent of 19th-century carpet sweepers by Bissell or Herrick. Robotic vacuums are now one of the fastest growing areas of the vacuum category, in part because they promise to clean for you. However, tests have confirmed that they are not nearly as powerful or effective as upright or cannister vacuums that must be plugged in.[30] Their batteries are too small to produce comparable suction power and drive a carpet sweeping brush alongside autonomous control and forward propulsion. If anything, feature expansion has moved towards secondary bells and whistles that consume more energy, such as voice control, security cameras and more, not improved cleaning performance. While a few robotic vacuum companies have claimed their products save energy, other studies have made contradicting assertions.[31] Today, traditional carpet sweepers and brooms remain capable tools in a home cleaning arsenal without using any energy at all.

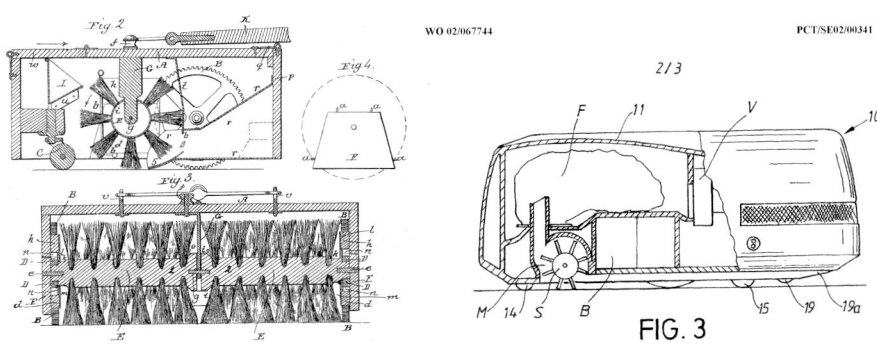

Figure 5.16
Above left: Hiram Herrick's 1859 patent for a push carpet sweeper that was later developed by Melville Bissel a few decades later. Source: USPTO.

Figure 5.17
Above right: A patent image of a 1990s Electrolux Trilobyte robotic vacuum. Notice the carpet sweeping component at the left of the illustration. In this case the sweeper just loosens dust, unlike the early model that collects it as well.
Source: WIPO.

Figure 5.18

A montage of carpet sweeper concepts, some from the author's industrial design studio—all considered renewable materials, aesthetics, durability and the home context. Students tested a recent and 100-year-old Bissell sweeper with users. Clockwise from top left: Andrew Ferrier's 2015 "Brush" concept; Oriana Nordt's renewable beechwood design stores and blends into home environments when not in use while making the emptying process easier for the user. Soren Winistorfer's beechwood sweeper blends into home environments to be left out. Katie LeMay's Flamingo sweeper reaches under furniture. Danielle Fattibene's toy inspired aesthetic encourage children to learn to tidy up. Reid Holbert's design explores improving the dust emptying process and RJ Weaver's employs "high tech" aesthetics and a planetary gear to increase sweeping performance. Design for disassembly, recyclability and user replacement of worn brushes were also considered in these designs and others. Source: Andrew Ferrier, Oriana Nordt, Katie LeMay, Soren Winistorfer, RJ Weaver & Reid Holbert.

Manual Push Lawn Mowers

In many respects, the development of the push lawn mower followed a similar trajectory to the manual carpet sweeper. The first lawn mower, pushed by hand, was invented by Edwin Budding in 1830 in the UK.[32] Budding's push mower was designed principally to cut the grass on sports fields and formal gardens, as a labor-saving alternative to the scythe. In 1830 it was granted a British patent.[33] Soon afterward, additional patents were issued in the US. By the early 1900s, gasoline-powered mowers were introduced and slowly became the ubiquitous tool for home lawn maintenance that they are today. Later on, by the early 21st century, as many as five million were sold per year in the US. Recent studies suggest that gasoline-powered lawn equipment can produce more air pollution and carbon emissions per hour than driving a car.[34] To add to these disadvantages, gas mowers also require engine maintenance—motor oil changes, tuning and spark plug replacement. Many models also produce considerable noise, with potential negative effects on hearing. Fortunately, electric mowers were introduced by the mid-2010s, which offer easier maintenance, less noise and at least no local air pollution during operation. Robotic variations of electric mowers, like their vacuum siblings, also emerged about the same time. At present, like all battery-electric appliances, electric mowers use large disposable lithium-ion batteries. While some recycling infrastructure exists, and while technologists are exploring how to make them recyclable, today lithium-ion batteries usually end up in landfills where they can leak and contaminate ground water.[35] These problems also plague other forms of battery-electric appliances from scooters to electric cars.

Interestingly, slightly before the electric mower was introduced, manual push mowers experienced a renaissance in the early 2000s. Many explanations have been offered for this phenomenon: they are lighter than gasoline (and electric-powered) models, they offer an opportunity for some light physical activity that one might otherwise get at a gym. In the case of gas mowers, they also don't require pull cord starting which some find unpleasant.[36] Others have suggested that customers who are concerned about the environmental impact of gas mowers drove their resurgence. Whatever the explanation, they can work well for homes with small lawns. Today more modern push mowers have reached the market using lighter materials and superior blades. Companies with expertise in making blades, scissors and shears, like Finnish manufacturer Fiskars, now offer these kinds of lawn mowers in the marketplace.

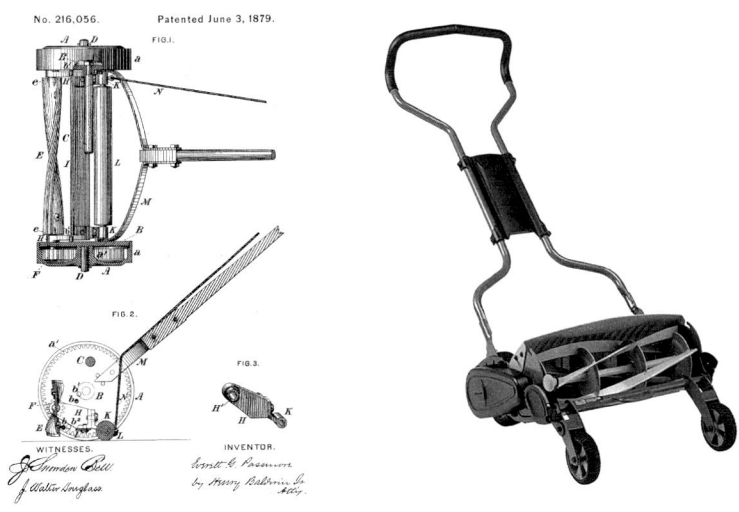

Figure 5.19
Above left: An 1879 patent for a push lawnmower. In the 20th century, gas-powered mowers replaced manual models. More recently manufacturers have introduced electric lawn care tools including mowers. Source: USPTO.

Figure 5.20
Above right: More recently, manual push mowers have seen a resurgence in popularity, especially for home owners with small lawns seeking low maintenance, environmental and modest exercise benefits. Manufacturers such as Fiskars developed this modern "Staysharp" series which offers a lightweight alternative to the cast iron originals. Source: Fiskars Group US.

Lawn Care Then and Now

Beyond the question of which mower to use today (a push mower, gas-powered or battery-electric), one would be remiss not to ask why having a grass lawn is such a non-negotiable feature of Anglo-Western culture in the first place. Indeed, a manicured grass lawn is an attractive amenity and perhaps one of the most identifiable components of the Western suburban landscape—a nod to the natural world and a symbol of middle-class achievement. In ideal circumstances, cropped grass lawns provide a place to relax, engage in sports and spend weekend time with friends and family. They are universally loved and dutifully cared for. According to a 2005 study, there are more than 40 million acres of grass lawns in the US between houses, parks and corporate campuses—more landmass than that of any irrigated crop.[37] Each year, roughly three trillion gallons of water are used to keep them green; 200 million gallons of gasoline are used to mow and maintain them; and 70 million pounds of pesticides are then used to keep them weed-free.[38]

Despite their ubiquity in the US, grass lawns are not native to the continent. Many grass species, including Kentucky Blue Grass, originated in parts of Asia, Africa and Europe but have thrived in areas of the US with sufficient rainfall. The popularity of lawns grew incrementally, beginning in the Middle Ages. Some etymologists believe the Middle English word *launde* originally described an opening in the forest, but later described artificial openings or glades that were cleared by humans.[39] Some of the earliest lawns were grass fields around castles in France and Britain, kept free of trees to enable soldiers and guards to see approaching, perhaps hostile, visitors. Maintaining these "lawns" was a matter of immense human effort and resources. Cutting and weeding these grassy fields was often performed by hand or with hand tools.

The term "lawn" later referred to the village "commons," or the meadows shared or held "in common" where villagers could let their livestock graze. These four-legged lawn mowers kept the grass clipped, while fertilizing as they ate. Of course, today "grass-fed beef" refers to the image of cattle grazing freely in an open field. While this has become an anachronism in beef production today, many countries have continued to use cattle, dairy cows and other animals to keep grasslands trimmed, often with a positive effect on the flavor of the meat and dairy products these animals produce. Mountainous countries in the Alps, for example, continue to rely on cows to climb steeper terrain and crop their sloped fields.

The contemporary popularity of the lawn spread with the British Empire and evolved into a global phenomenon in tandem with a number of lawn-based sports, such as cricket and football (soccer). The lawn's popularity in the US as a public and middle-class amenity is likely derived from European tastes, particularly that of the English. George Washington and Thomas Jefferson both had extensive manicured lawns at their family estates and relied on slave labor for their upkeep. The architecture and grounds of these estates were, by design, based on neo-classical and European precedents and likely had influence on the broader American taste for grass lawns that followed. Later on, Frederick Law Olmstead, the pre-eminent American landscape designer of New York City's Central Park and many other notable public green spaces and developments, believed lawns were a vital feature of the public park, offering needed respite from industrializing cities in the mid to late 19th century.

As the presence of lawns has become entrenched, so too has their significant environmental cost—in addition to the emissions and energy consumption needed by powered lawn mowers, lawns also need generous watering to resemble the green, lush, bucolic archetypes on which they are based. Many areas of the arid western US and other parts of the world adopted lawns even as water supplies have diminished. Today, in response to several consecutive years of drought, California, Nevada and other states of the Rocky Mountains and southwest have offered residents financial incentives to remove grass lawns in favor of less thirsty alternatives.[40]

The No-Mow Movement and the "Paris Lawn"

In parallel to water conservation efforts in US southwest, a broader "No-Mow" movement has emerged which offers aesthetic design alternatives to the grass lawn. The National Resource Defense Council, an environmental advocacy group, defines four categories within this definition: 1) naturalized or un-mowed turf grass that is left to grow wild; 2) low-growing turf grasses that require little grooming (most are a blend of fescues); 3) native or naturalized landscapes where turf is replaced with native plants as well as noninvasive, climate-friendly ones that can thrive in local conditions; and 4) yards where edible plants grow—vegetables and fruit-bearing trees and shrubs.[41]

It is questionable whether the majority of Americans would voluntarily give up their lawns to a wild, native alternative. Local town ordinances also prohibit untrimmed grass lawns, reasoning that they attract pests, diminish real estate values and are unneighborly.[42] On the other hand, it might be possible to find a compromise between the trimmed aesthetic of the lawn and one that uses less resources. One example is the "Paris Lawn," named after the Paris Climate Agreement. In this proposed landscape design, a large section of a back yard's grass is left to grow untrimmed through the summer or grow season. By cutting it into a uniform shape, like a square (perhaps 20' x 20' but adjusted to the yard's size) it can preserve a manicured, landscaped effect. For those who have adopted robotic mowers, they could even be programmed or "hacked" to produce decorative, custom outlines for visual expression.[43]

Further Water Consumption in the Home

At present, nearly two-thirds of the world's population faces some form of water scarcity for drinking, agriculture and sanitation purposes.[44] At the same time, wealthier, industrially developed countries use far more water per capita than their companion developing nations.[45] As recently discussed, grass lawns account for a significant share of home water consumption. Additionally, modern appliances including showers, flush toilets and washing machines have also contributed to increasing water consumption in industrialized economies despite their improved sanitation and convenience benefits. Advances in the appliances themselves, including low flow faucets, shower heads and toilets, have provided notable savings in water consumption. What other kinds of water saving strategies could be used to conserve water in the home?

Water Collection: Rain Cisterns

Water supply into most Western homes relies on municipal sources which are tapped from rivers, aquifers and other fresh water sources and then treated for reuse. In the US, 15% of homes use private ground wells, particularly in rural and agricultural regions.[46] Over time, in many areas, the water table has receded; in other words, wells have to reach deeper to access groundwater. Chicago's ground water, for example, has fallen to 900feet below the surface according to the US Geological Survey study.[47]

Collection of rainwater is a practical, if incomplete, alternative way to provide water, dating back thousands of years to Greece,[48,49] Mesopotamia, China, India and the Americas. In addition, it was commonly used by European settlers to North America in the 17th and 18th centuries. Of course, rain has to be plentiful to be worthwhile. Roofs on buildings were specifically designed to divert rainwater into cisterns for drinking, agriculture and sanitation purposes. After use of these systems declined during a period of urbanization in the 19th century, rain cisterns have seen a revival for their simplicity and efficacy. Organizations like the American Rainwater Catchment Systems Association (ARCSA) have endeavored, in their words, to:

> bring renewed attention to the ancient practice of rainwater harvesting. For thousands of years collecting rainwater was a common method for providing water, but over the last century, wells and municipal water supplies took over as primary water sources. The diminishing supply of fresh water in wells and aquifers, concerns of quality and population growth are among the top reasons for the resurgence of rainwater catchment.[50]

Yet, despite the promise of rainwater collection, most of the designs for these systems are an afterthought to the architecture and homes they support. How can design be used to reintegrate roof top water capture back into buildings themselves?

In projects today, Iran-based BMDS Architects designed a bowl-shaped rooftop rainwater collector that looks equally sculptural as it is utilitarian. Netherlands-based Elho developed the Pure Raindrop water barrel for existing homes. Based on designer Bas Van de Veer's design school graduation project, this iconic, user-friendly system integrates into existing roof drain pipes like conventional rain barrels but with elegant design and recyclable materials. Numerous other intriguing projects have been proposed by designers and design firms including Rua Acquitectos in Brazil and Shaakira Jassat from South Africa and Genesis Solano in the US.

Figure 5.21
Above left: The Concave Roof System rainwater collector for modern homes by Iran-based BMDS Architects. Source: BM Design Studios.

Figure 5.22
Above right: A Pure Raindrop rainwater cistern designed by Bas Van de Veer and made by Netherlands-based Elho. The 70liter container connects to a home's downspout to collect water for plants and home gardening. Source: Studio Bas van der Veer.

Fog Water Harvesting

Perhaps more obscure but gaining attention, fog and dew harvesting are older practices of converting fog or condensation into usable, drinkable water. Fog is composed of small water droplets suspended in the air which are formed in different climactic conditions around the world, often in arid coastal regions. Archaeological evidence has shown that human settlements used various forms of the technology dating back thousands of years.[51] Stone structures in Israel, Iran, Peru, Egypt, England and some Mediterranean islands were constructed to collect fog water for irrigation where natural water sources were limited. Sometimes arranged in a circular plan, fog and dew collectors would collect on stone walls and drip down to the soil below where plants would be sustained to grow and thrive. Many of these technologies have seen a few adaptations for contemporary use.

On the Italian island of Pantelleria, there is a unique native structure called a Giardini Panteschi, or Pantellerian Garden. Built of local volcanic stone, the surface of the circular structure gathers fog that passes over the windy island and distributes it to the trees that are planted inside the wall. The top of the wall slopes inward allowing collected water to be drawn inward to moisten the soil where often citrus fruit trees are planted. The wall is also built to a precise height to simultaneously protect younger trees from wind exposure while allowing ample sunlight in throughout the year. Additionally, the volcanic rock collects heat from the sun during the day, which radiates heat for the plants overnight. A similar technique, known as a "Fruit Wall," has been used for hundreds of years in western Europe to support more tropical plants in northern climates. Elements of the design of Giardini Panteschi overlap with the designs of similar structures around the world.[52]

In addition to these stone structures, plants themselves have been used to collect fog droplets. The coastal redwoods of Northern California (*Sequoia Sempervirens*) and similar trees in Hawaii and other continents engage in a natural phenomenon called "Fog Drip" whereby droplets collect on the pine nettles and drop to the forest floor where they are absorbed for use by the tree or other plants in the ecosystem. Sequoia trees have the remarkable ability to satisfy roughly 34% of their water needs through passing fog alone.[53] Other plants have been shown to do the same. A native Chilean bush called *Tillandsia*, found in the highly arid Atacama Desert, is also capable of collecting passing fog.[54] Cacti and other plants have likewise been studied in the broad developing field of bio-inspired design for similar reasons. Like many of the alternative and traditional technologies investigated in this book, nature itself provides a valuable resource of alternative design and technology aimed at achieving similar, more sustainable, outcomes. Numerous volumes now summarize this burgeoning field, including works such as Janine Benyus' *Biomimicry: Innovation Inspired by Nature* and Michael Palwyn's *Biomimicry in Architecture*.[55]

Additional archaeological evidence demonstrates that plants and shrubs were used to collect water hundreds of years ago by simply constructing a dish or pond underneath the plant to gather dripping water. Other precedents have been found in Peru, built by the Inca where buckets were placed below trees. In 2009, German researchers Kai Tiedemann and Anne Lummerich planted 800 she-oak trees in Peru to create a natural fog-catching system that aimed to replicate this ancient technique.[56] In the Canary Islands, off the coast of Senegal, a native species of Laurel tree (Garoé or Ocotea Foetes) has been used for hundreds of years to collect fog water for agriculture. Even today, forests of these "Fountain Trees" continue to be used to provide drinking water for goats and sheep on the westerly island of Hierro. So famous are these trees to the local population that a symbol of the Fountain Tree is emblazoned on the island's flag.

Figure 5.23
Above left: A c.1605 drawing by Claude Duret of a "water tree" (*Un arbre porte-eaux*) used to capture drinking water on the Canary Islands.[57] Source: Claude Duret.

Figure 5.24
Above right: Fog Harp, a tree-inspired fog-harvesting concept. Source: Brook Kennedy.

At present, fog harvesting is drawing considerable attention for its potential to provide fresh water to communities suffering the greatest from water scarcity. Today, more than four billion people face water scarcity at least one month per year and these figures are expected to grow.[58] For communities that lack municipal water, fog harvesting can replace or supplement well water if fog is present. In 1969, one of the first modern fog collectors was tested at a navy base in South Africa and later, in 1987, Canadian NGO Fog Quest developed a similar structure for use in northern Chile.[59] These modest devices, loosely resembling volley ball nets, rely on low-cost knitted plastic mesh materials, often "Raschel" mesh. While this material is functional and collects fog water, recent material research science and design efforts have been making progress towards increasing the water yield of such devices.[60] Designers Peter Trautwein of Aqualonis and Arturo Vittorio of Warka Water have been working towards improving their yield and acceptance. Frequent maintenance, cultural perceptions and other issues pose barriers to the widespread adoption of these devices. Even so, for many less affluent countries, women and children can spend hours per day fetching and retrieving water for their families, taking valuable time away from education and other productive tasks.[61] Wealthier nations, such as the United Arab Emirates and South Africa, have explored fog harvesting on a larger scale for drinking, industry and to reverse the effects of desertification. California's agricultural regions have also begun to explore fog harvesting, especially since the onset of the recent multi-year drought where local rivers, like the mighty Colorado, have lost significant volume.

Sea water desalination is a relatively new water technology which is now used in some more populous global regions lacking sufficient water. It is gaining ground in part because of the large volume of water it can provide. San Diego, California now obtains roughly 10% of its water from desalination to supplement declining supplies from other sources. While solar-electric methods have been tested with desalination, this technology remains energy intensive. Fog harvesting, in contrast, is passive (although electrically enhanced variants are being tested as well). Generally, the technology is best suited to remote homes and communities where no other water supply exists. Nevertheless, all of these methods for collecting alternative sources of water are essential as supply disappears. Collectively, they could all play a role in global water management strategies alongside water reuse and efforts to reduce consumption.

Symbiotic Upkeep: Harboring Animals for Mutual Benefit

Just as trees and other structures have been useful for passively collecting drinkable water for homes and communities, harboring animals have also served tasks that are similarly relevant today. Common pests, for instance, can plague crop storage in agricultural barns and home gardens. While traps and poisons are often used today, owls can also be enlisted as natural predators. Through architectural design, an entrance and sometimes a landing platform high up on a barn's gable can encourage owls to move in and serve this func-tion. These *Owl Holes* were first seen in the design of barns in western Europe and North America, but few survive today. In a similar vein, *Dovecotes* provided shelter for birds in trade for their eggs, fertilizing dung and sometimes their services to bring messages over long distances. Examples of these architectural features can be found in England, Iran and the US. Today, conceptually related designs like *BeeBrick* provide a secure habitat for pol-linating bees which in turn help support local agriculture and gardens.[62] Perhaps there are other tasks that might be possible by carefully integrating safe natural habitats for other liv-ing organisms into our homes. These kinds of symbiotic relationships have to be considered carefully to avoid unintended consequences but could provide some worthwhile value to human and animal livelihoods.

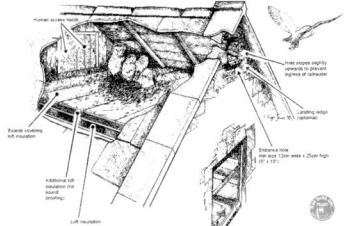 

Figure 5.25
Above left: A schematic illustration showing how owls can inhabit the loft areas of barns. These holes invite barn owls to roost inside where they can manage rodent populations without using disposable traps or harsh chemicals. Source: The Barn Owl Trust, www.barnowltrust.org.uk.

Figure 5.26
Above right: A Beebrick bee habitat integrated into a building's brick wall helps support declining bee populations. Source: Green&Blue.

Notes

1 Fanara, A., Clark, R., Duff, R. and Polad, M. (2007) 'How small devices are having a big impact on U.S. Utility bills,' *Energy Star (US Environmental Protection Agency & US Department of Energy)* [online]. Available at: www.energystar.gov/ia/partners/prod_development/downloads/EEDAL-145.pdf (Accessed: August 3, 2022).

2 The Good Housekeeping Institute (2015) 'At home in America: A snapshot of the average household in 2015,' *Good Housekeeping*. Article Appeared Originally in the November, 2015 Good Housekeeping Magazine, October 19 [online]. Available at: www.goodhousekeeping.com/home/g2864/american-home-survey-2015/ (Accessed: August 9, 2022).

3 Fullerton, H. (1999) 'Labor force participation: 75 years of change, 1950–98 and 1998–2025,' *Monthly Labor Review* [online]. Available at: www.bls.gov/opub/mlr/1999/12/art1full.pdf (Accessed: August 9, 2022).

4 Strasser, S. (1982) *Never done: A history of American housework*. New York: Pantheon Books.

5 Cowan, R.S. (1985) *More work for mother: The ironies of household technology from the open hearth to the microwave*. New York: Basic Books.

6 US EIA (2019) *Use of energy explained. Energy use in homes* [online]. Available at: www.eia.gov/energyexplained/use-of-energy/electricity-use-in-homes.php (Accessed: August 9, 2022).

7 Minds, S. (n.d.) '16 ways to save money in the laundry room,' *Energy Saver* [online]. Available at: www.energy.gov/energysaver/articles/16-ways-save-money-laundry-room (Accessed: December 7, 2022).

8 Pentland, W. (2013) 'Europe's clothes dryers consume half as much energy as America's,' *Forbes*, June 11 [online]. Available at: www.forbes.com/sites/williampentland/2013/06/11/europes-clothes-dryers-consume-half-as-much-energy-as-americas/?sh=675b4a935e67 (Accessed: December 7, 2022).

9 Gluesenkamp, K.R., Shen, B. and Boudreaux, P. (2017) 'Energy efficient clothes dryer—final report,' *Oak Ridge National Laboratory*, August 31 [online]. Available at: https://info.ornl.gov/sites/publications/Files/Pub139484.pdf (Accessed: December 9, 2022).

10 Casey, T. (2022) 'Suddenly, heat pump clothes dryers are having a moment: The energy efficient heat pump clothes dryer is finally ready for its closeup,' *CleanTechnica*, April 4 [online]. Available at: https://cleantechnica.com/2022/04/04/suddenly-heat-pump-clothes-dryers-are-having-a-moment/ (Accessed: August 9, 2022).

11 Hughes, K.A. (2007) 'To fight global warming, some hang a clothesline,' *The New York Times*, April 12 [online]. Available at: www.nytimes.com/2007/04/12/garden/12clothesline.html (Accessed: August 9, 2022).

12 Sweets Catalogue (1910) 'Sweets catalogue of building construction for the year 1910,' p1082. Available at: https://babel.hathitrust.org/cgi/pt?id=uiuo.ark:/13960/t3710nk4j&view=1up&seq=1172&skin=2021&q1=chicago%20clothes%20dryer

13 Kohlstedt, K. (2017) 'Hills hoist: Iconic rotary clothesline that shaped suburban Australia,' *99% Invisible*, May 1 [online]. Available at: https://99percentinvisible.org/article/hills-hoist-iconic-rotary-clothesline-shaped-suburban-australia/ (Accessed: August 9, 2022).

14 Woodside, C. (2017) 'Drawing a line on outdoor clothes drying,' *The New York Times*, December 2 [online]. Available at: www.nytimes.com/2007/12/02/nyregion/nyregionspecial2/02clotheslinect.html (Accessed: August 9, 2022).

15 US Environmental Protection Agency (2018) *Nondurable goods: Product-specific data* [online]. Available at: www.epa.gov/facts-and-figures-about-materials-waste-and-recycling/nondurable-goods-product-specific-data#tab-6 (Accessed: August 9, 2022).

16 Greiling, K. (n.d.) *Extra Hands* [online]. Available at: www.studiogreiling.com/EXTRA-HANDS-FURNISHING-UTOPIA (Accessed: December 20, 2022).

17 Chamberlin et al. (1939) *Cleaning textile and similar materials*. US2165884.

18 De Decker, K. (2014) 'If we insulate our houses, why not our cooking pots?' *Lowtech Magazine* [online]. Available at: www.lowtechmagazine.com/2014/07/cooking-pot-insulation-key-to-sustainable-cooking.html (Accessed: August 9, 2022).

19 Cushing, F.H. (1883) 'A study of Pueblo pottery as illustrative of Zuñi Culture-Growth. Fourth Annual Report of the Bureau of Ethnology to the Secretary of the Smithsonian Institution 1882–1883,' *Smithsonian Institution. Bureau of Ethnology*, pp473–521. Available at: https://archive.org/details/annualreportofbu418821883smit/page/486/mode/2up (Accessed: August 9, 2022).

20 Scientific American (1869) 'The self-acting Norwegian cooking apparatus,' *Scientific American*, 21(11), p161. www.jstor.org/stable/26034508 (Accessed: August 9, 2022).

21 De Decker (2014).

22 Marsh, E. (2017) *The Fireless Cooker*. Available at: www.nal.usda.gov/collections/stories/fireless-cooker (Accessed: August 9, 2022).

23 Nummer, B.A. (2002) 'Historical origins of food preservation,' *National Center for Home Food Preservation*. Available at: https://nchfp.uga.edu/publications/nchfp/factsheets/food_pres_hist.html (Accessed: August 9, 2022).

24 Evans, L. (2000) 'The advent of mechanical refrigeration alters daily life and national economies throughout the world,' *Science and Its Times*, 5, p537. ISBN 0-7876-3937-0. (Accessed: August 10, 2022).

25 Sweets Catalogue (1910) *C.H. Stephenson Manufacturer of the Stephenson Sanitary Household Specialties* [online], pp850–851. Available at: https://babel.hathitrust.org/cgi/pt?id=uiuo.ark:/13960/t3710nk4j&view=1up&seq=942&skin=2021&size=125&q1=garbage (Accessed: August 9, 2022).

26 Herrick, H.H. (1858) *Carpet sweeper* [online]. Available at: https://patents.google.com/patent/US21233A/en (Accessed: August 9, 2022).

27 Bissell, M.R. (1876) *Carpet sweeper* [online]. Available at: https://pdfpiw.uspto.gov/.piw?PageNum=0&docid=00182346 (Accessed: August 9, 2022).

28 Gantz, C. (2012) *The vacuum cleaner: A history*. Jefferson: McFarland and Co.

29 GM's newest concept car, the Cadillac Celestiq, is similarly presented as an expression of the Zeitgeist. Carpet sweepers could similarly be designed to embrace sustainable design objects articulated by the UN Sustainable Development Goals.

30 Rae, H. (2020) 'Can a robotic vacuum replace your upright?' *Consumer Reports*, November 18 [online]. Available at: www.consumerreports.org/robotic-vacuums/can-a-robotic-vacuum-replace-your-canister-or-upright-a1557864633/ (Accessed: August 9, 2022).

31 Nicholls, L. and Strengers, Y. (2019) 'Robotic vacuum cleaners save energy? Raising cleanliness conventions and energy demand in Australian households with smart home technologies,' *Energy Research & Social Science*, 50, pp73–81. https://doi.org/10.1016/j.erss.2018.11.019.

32 Radam, B. (2020) *Lawnmowers: An illustrated history*. Stroud: Amberley Publishing.

33 Passmore, E.G. (1879) *Improvement in lawn-mowers* [online]. Available at: US RE 8560 (Accessed: August 9, 2022).

34 California Air Resources Board (2017) *Small engines in California* [online]. Available at: www.arb.ca.gov/msprog/offroad/sore/sm_en_fs.pdf?_ga=2.71233360.1364124981.1624629871-1659041529.1623334947 (Accessed: August 9, 2022).

35 Jacoby, M. (2019) 'It's time to get serious about recycling lithium-ion batteries,' *Chemical and Engineering News*, 97(28).

36 Csepiga, M. (2007) '"Reel" resurgence for old-fashioned mowers,' *Northwest Indiana Times*, August 10 [online]. Available at: www.nwitimes.com/business/local/reel-resurgence-for-old-fashioned-mowers/article_07befd40-c233-5ff0-b6a4-b754c55d01b1.html

37 Milesi, C., Running, S.W., Elvidge, C.D. et al. (2005) 'Mapping and modeling the biogeochemical cycling of turf grasses in the United States,' *Environmental Management*, 36, pp426–438.

38 Talbot, M. (2016) 'More sustainable (and beautiful) alternatives to grass,' *NRDC* [online]. Available at: www.nrdc.org/stories/more-sustainable-and-beautiful-alternatives-grass-lawn (Accessed: August 9, 2022).

39 Middle English Compendium (n.d.) *Launde*. Available at: https://quod.lib.umich.edu/m/middle-english-dictionary/dictionary/MED24859

40 Zhang, X. and Khachatryan, H. (2020) 'Investigating monetary incentives for environmentally friendly residential landscapes,' *Water*, 12(11), 3023.

41 Talbot, M. (2016) 'More sustainable (and beautiful) alternatives to a grass lawn,' *NRDC*, September 30 [online]. Available at: www.nrdc.org/stories/more-sustainable-and-beautiful-alternatives-grass-lawn (Accessible at: August 9, 2022).

42 Denvir, A., Meehan, M., Pellegrino, M. and Pratt, L. (2016) 'Towards sustainable landscapes: Restoring the right NOT to mow,' *Environmental Protection Clinic* [online]. Available at: www.nrdc.org/sites/default/files/sustainable-landscapes-20160506.pdf [Accessed: August 9, 2022).

43 D'Costa, K. (2017) 'The American obsession with lawns,' *Scientific American* [online]. Available at: https://blogs.scientificamerican.com/anthropology-in-practice/the-american-obsession-with-lawns/ (Accessed: August 9, 2022).

44 Mekonnen, M.M. and Hoekstra, A.Y. (2016) 'Four billion people facing severe water scarcity,' *Sci. Adv*, 2(2). Available at: https://doi.org/10.1126/sciadv.1500323 (Accessed: August 9, 2022).

45 Water Footprint Calculator (2017) *Water footprint comparisons by country* [online]. www.watercalculator.org/footprint/water-footprints-by-country/ (Accessed: August 9, 2022).

46 Water Resources (2019) 'Domestic (private) supply wells,' *US Geological Survey* [online]. Available at: www.usgs.gov/mission-areas/water-resources/science/domestic-private-supply-wells (Accessed: August 9, 2022).

47 Alley, W., Reilly, T.E. and Franke, O.L. (1999) *Sustainability of ground-water resources. U.S. Geological Survey*. Available at: www.usgs.gov/media/images/map-chicago-milwaukee-area-showing-water-level-decline-1864-1980 (Accessed: August 9, 2022).

48 Sazakli, E., Sazaklie, E. and Leotsinidis, M. (2015) 'Rainwater harvesting: From ancient Greeks to modern times. The case of Kefalonia Island,' *International Journal of Global Environmental Issues*, 14, pp286–295. DOI:10.1504/IJGENVI.2015.071867.

49 Mays, L., Antoniou, G. and Angelakis, A. (2013) 'History of water cisterns: Legacies and lessons,' *Water*, 5(4), pp1916–1940. DOI:10.3390/w5041916.

50 American Rainwater Catchment Systems Association (ARCSA) (n.d.) *History and background* [online]. Available at: www.arcsa.org/page/arcsa_history (Accessed: August 9, 2022).

51 Gioda, A., Hernández, Z., Gonzáles, E. and Espejo, R. (1995) 'Fountain trees of the Canary Islands: Legend and reality,' *Advances in Horticultural Science*, 9(3) pp112–118.

52 Atlas Obscura (n.d.) *Giardino Pantesco*. Available at: www.atlasobscura.com/places/giardino-pantesco-of-pantelleria (Accessed: March 23, 2021).

53 Dawson, T. (1998) 'Fog in the California redwood forest: Ecosystem inputs and use by plants,' *Oecologia*, 117, pp476–485. DOI:10.1007/s004420050683

54 Amos, J. (2015) 'Fog history of Atacama reconstructed,' *BBC*, December 21. Available at: www.bbc.com/news/science-environment-35065404 (Accessed: March 23, 2021).

55 Benyus, J. (1997) *Biomimicry: Innovation inspired by nature*. New York: William Morrow; Palwyn, M. (2016) *Biomimicry in Architecture*. London: RIBA Publishing.

56 Tiedemann, K. and Lummerich, A. (2010) 'Fog harvesting on the verge of economic competitiveness,' *Erdkunde*, 65(3). DOI: 10.2307/23069701

57 Duret, C. (1605) *Histoire Admirable des Plantes et des Herbes Esemerveillables*. Paris: Nicolas Buon, p. 209.

58 Mekonnen and Hoekstra (2016).

59 Fessehaye, M., Abdul-Wahab, S.A., Savage, M.J., Kohler, T., Gherezghiher, T. and Hurni, H. (2014) 'Fog-water collection for community use,' *Renewable and Sustainable Energy Reviews*, 29, pp52–62. DOI:10.1016/j.rser.2013.08.063

60 The author discloses that he worked on fog harvesting technology prior to releasing this book.

61 Trevino, M. (2020) 'The ethereal art of fog-catching,' *BBC Future Planet*, February 23 [online]. Available at: www.bbc.com/future/article/20200221-how-fog-can-solve-water-shortage-from-climate-change-in-peru (Accessed: August 9, 2022).

62 United Nations Food and Agriculture Organization (2018) 'Why bees matter,' May 20 [online]. Available at: www.fao.org/3/i9527en/i9527en.pdf (Accessed: August 9, 2022).

Return to Durable Design

The rise of disposable design has shaped much of the legacy of industrial design since the mid to 20th century as single-use products have infiltrated the marketplace—PET plastic water bottles and takeaway food packaging are commonly cited examples. Tragically, many categories of disposable products evolved from durable precedents. Durable goods are generally defined as products that are made to last for a long time, at least a few years.[1] Today, they are frequently cited as a vital part of transitioning to a *Circular Economy* to conserve natural resources and to help eliminate pollution and landfilled waste.[2] Durable goods include many categories: automobiles, major home appliances, furniture, medical equipment and jewelry. Disposable goods (or non-durables) include single-use beverage containers, disposable take-away packaging but also short-lived goods like razor cartridges that are disposed after only a few weeks of use. Often disposable goods are not recyclable. Durable goods, on the other hand, are frequently made of sturdier materials that are able to last longer, be refurbished or repaired when components break or wear out. While much recent ecological thinking has emphasized recyclability and upcycling, other researchers have begun to emphasize the value of durability. For instance, sustainable design expert Tim Cooper has long advocated that recycling derelict or used products is an insufficient strategy—in addition, Cooper believes that more emphasis should be placed on reducing consumption, in part by making products last longer.[3] A component of this approach to durable goods is that owners are sometimes able to repair these products themselves. However, since the second half of the 20th century, durable products of all kinds gradually became harder to repair, especially as they have become more complex. This is not necessarily an accident—commonly, this phenomenon is called "planned obsolescence" where consumers are pressured to buy a replacement product when the one they have breaks prematurely or cannot bo fixod or upgraded.[4,5] In this chapter, an argument will be made in favor of resuscitating durable design, a practice which has been gaining new momentum in recent years. Today, more than ever, it deserves to be rediscovered and advanced in the century ahead to mitigate ever increasing volumes of landfilled waste from the built environment.

As the European Consumer Organization (BEUC) reports, it is a common occurrence that durable goods break prematurely because critical components of their design fail, without the possibility of repair.[6] The author experienced such an event with an electric water kettle. After just three years of daily use, a small brass connector (which connects the kettle to the power source which heats the water) fractured. The failed part could neither be replaced nor accessed for removal and refurbishment. As a result, the manufacturer's customer service department recommended that the kettle be disposed and replaced. This outcome obviously serves the manufacturer's interests who can benefit from selling a new

DOI: 10.4324/9780367814304-7

one, but ultimately impacts society by filling landfills with tons of minimally worn products. Many of these products are also full of useful spare parts that could help repair other broken products. Sometimes, even unused returned products end up in landfills as well.[7]

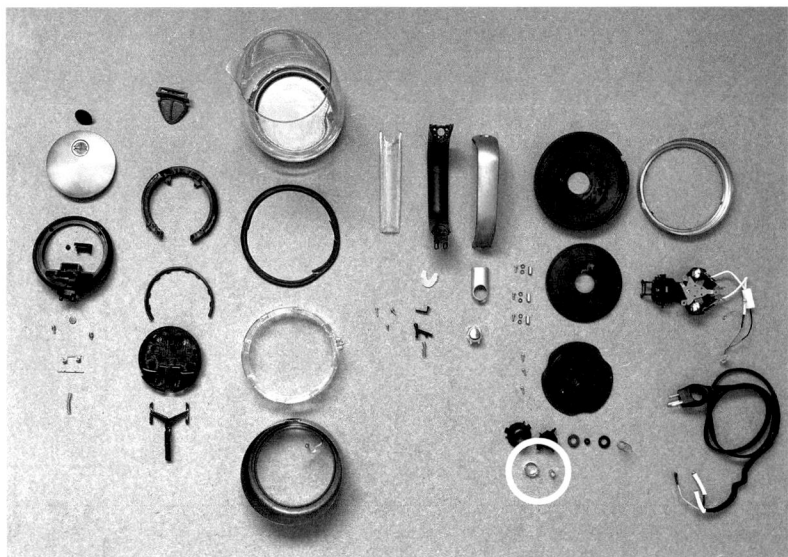

Figure 6.1
An unidentified electric water kettle taken apart in a design class, showing the component that broke leading to its disposal. Only after breaking the product housing open could the failed component be accessed. Source: Brook Kennedy.

In many other examples, products can become unusable when simple components break, often internal electronic components. Many leading brands of electric toothbrushes, including Philips Sonicare™, have long received favorable reviews by dentists for their hygienic performance compared to analog alternatives. Yet, unlike analog bristle toothbrushes used for hundreds of years prior to electrification,[8] many electric toothbrushes rely on rechargeable lithium-ion batteries just like those used in many electric products from smartphones to electric cars. Here, they are often charged inductively, in part to insulate the electrical connection from wet bathroom environments. As a result, when the batteries cease to hold a charge (an inevitable result of their use in any rechargeable product), users are unable to replace them nor are there facilities that specialize in their repair. Thus, they are thrown away and the batteries themselves become hazardous waste in landfills when their specialized chemistries leach into groundwater. Apple and other smartphone manufacturers at least offer self-repair kits and self-help videos, but most users rely on professional battery replacement services. Even so, in the US, as many as 151 million phones are thrown away each year when they could otherwise be repaired.[9]

Other products can be opened and repaired but their manufacturers often don't want them to be.[10] In contrast, Fairphone, a Netherlands-based mobile phone manufacturer, has built a business on making their phones intuitive for their owners to repair and modify. Other durable goods' manufacturers have also begun to encourage the age-old value of repair or offer extended warranties for their refurbishment.

Figure 6.2
A 2021 Fairphone 4. Fairphone offers a completely repairable smartphone—users are able to open and replace the battery, cameras, audio units and more when they wear out or break without a technician or special tools. Source: Fairphone.

Right to Repair or Design for Maintainability (and Upgradability)

Legislation has also been advanced in several countries to require manufacturers to embrace repairability. In November 2020, the European Union successfully voted to adopt more aggressive measures to reduce electronic waste. The result might lead to broad legislation under the banner "Right to Repair."[11] Now laws have been enacted requiring companies to accept broken products or make institutional provisions for their repair. Similar legislation is being considered in the US in as many as 21 states, but with considerable opposition from industry.[12] In 2003, the US state of Massachusetts passed legislation protecting the right to repair automobiles. In 2022, right to repair legislation known as the "Digital Fair Repair Act" passed in the US State of New York, covering digital electronic equipment.[13] Unfortunately, the need to force manufacturers to make their products repairable through regulation was not always necessary. From automobiles, to shoes and home appliances, many products once were designed to be repaired and maintained by their owners or networks of technicians. More recently, some computers, like the Apple G4 Cube, were even designed to be upgraded. Faster processors, RAM and video cards could be easily replaced, thereby extending the computer's useful life. Today, some automobiles are upgraded with new engines, including contemporary electric drivetrains, to extend their lifespan.

Repair Infrastructure: Repairable Shoe Soles

Most shoes used to be designed to last for decades or even a lifetime. Whether boots, formal dress shoes or otherwise, shoes made from more durable (and precious) materials like

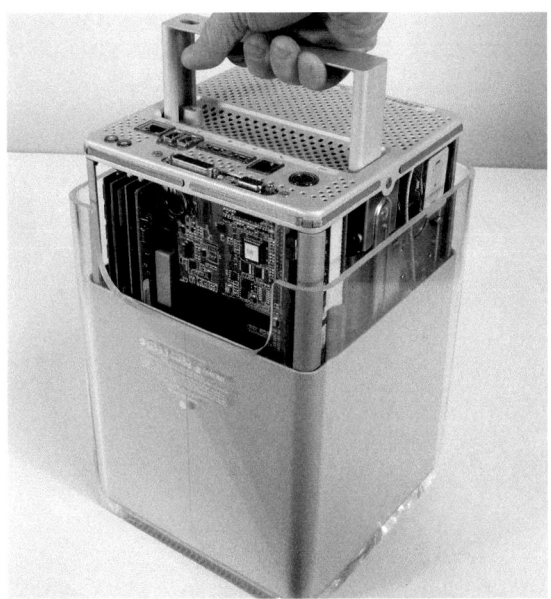

Figure 6.3
An Apple G4 Cube from 2001, by Johnny Ive, was deliberately designed to be upgraded. Like earlier Power Macs from the 90s the product's useful life could be extended by installing a newer processor and more memory. Source: Binarysequence (CC BY-SA 4.0).

leather and rubber could be re-polished, oiled and re-soled indefinitely. Their appearance, in many cases, could also improve with time and care. Cobblers or shoe repair shops in industrialized economies used to be ubiquitous and essential for maintaining shoes, although shining shoes can easily be done at home. Unfortunately, these repair facilities, once a neighborhood necessity, have declined as shoes are now designed to be harder to re-sole. Athletic shoes especially wear out quickly without options for repair.[14] Replacing a worn-out sole would certainly be far less wasteful than replacing a pair of shoes entirely. Furthermore, athletic shoes are often made of several types of foams and fabrics fused together such that they can't be recycled either when they wear out. What if shoe manufacturers were to design a line of boots, dress shoes and athletic shoes with soles that could be easily removed and replaced by the owner just as electronic components are with a Fairphone? Looking ahead, as society endeavors to reduce resource consumption and industrial waste, hopefully more manufacturers will offer ways to turn disposable products back into easily repairable durables. Until that happens, pioneering online repair services like iFixit and repair cafés have contributed volumes by helping a global community repair the broken durable goods around them.[15] Additionally, 3d printed replacement parts have also offered opportunities to extend the life of products through on-demand repair.[16]

Repair Infrastructure: Refurbished Furniture

Like buildings and railroad equipment, furniture can be renovated with fresh paint, wood finishes, upholstery and fabrics. Indeed, furniture can be rehabilitated by craftspeople one at a time. However, in some rare cases, manufacturers continue to offer refurbishment of the products they have mass-produced. For example, Swedish furniture maker Gärsnäs produced several hundred Riksdagen chairs for the Swedish Parliament in 1982, designed by Åke Axelsson. Embedded in the design were simple features and connectors that enabled the chair to be easily disassembled for refurbishment. In 2015, the well-used parliamentary chairs were shipped back to Gärsnäs for renewal with fresh upholstery, fabric and refinishing and are now back in daily use with a warranty to 2040.[17] Wood and upholstered furniture like the Riksdagen chair lend themselves to repair—plastic and metal framed chairs, on the other hand, can discolor and corrode over time and can't be as easily repainted or refinished to restore them to their original condition.

Design to Prevent Damage: The Swift Rise and Fall of the US 5mph Bumper

Beginning in the 1974 automotive model year, US market cars were required to be fitted with "5mph" bumpers. Based on regulations passed in 1971, the legislation was meant to reduce costly damage to cars hit at low speeds, 5mph or less. Some of these brawny bumpers were composed of a rubber crumple zone so that if hit by another car, the bumper could simply be pulled out again without visible damage. However, after 1990, the legislation was overturned. Automotive bumpers began to be replaced with more aerodynamic plastic shells with the intent of improving fuel economy (and aesthetics). Now, after even minor "bumps" or collisions, perhaps when parallel parking, present "bumpers" become scratched and chipped and are often fully replaced in insurance claims. As a consequence, these huge pieces of plastic are discarded. Could bumpers and other products, like smartphones, be designed to withstand minor bumps without showing unappealing damage? Smartphones and cars are supported by protective accessories, like cases and bumper protectors. Why can't these products simply be designed to prevent visible damage in the first place?

Emotional Durability: Designing Emotional Bonds Between Users and Products

Beyond the physical, utilitarian traits of products that support Circular Economy goals through longer product lifespans, researchers have also explored how to build lasting emotional bonds between users and products. Researcher Jonathan Chapman, for example, has written extensively about "emotionally durable design." Here Chapman theorizes that consumption and waste of natural resources can be reduced by increasing the "resilience of relationships established between consumers and products." Essentially, product replacement is delayed by developing strong emotional ties between a user and a product, similar in some respects to the kind of relationship that develops between people. In his book, *Emotionally Durable Design: Objects, Experiences & Empathy*[18], Chapman describes how:

> the process of consumption is, and has always been, motivated by complex emotional drivers, and is about far more than just the mindless purchasing of newer and shinier things; it is a journey towards the ideal or desired self, that through cyclical

loops of desire and disappointment, becomes a seemingly endless process of serial destruction.[19]

To avoid this undesirable outcome, "emotional durability" can be achieved through consideration of five elements: *Narrative*, *Consciousness*, *Attachment*, *Fiction* and *Surface*. Altogether, Chapman suggests that conscious design decisions can encourage an ever deeper emotional bond with a product over time. In the case of *Surface*, for example, materials can be chosen that develop more visual appeal rather than less over time, leading to a stronger personal human–product attachment. Blue jeans, for example, no matter how worn or tattered, often become more valued, rather than less valued, as they age and soften.[20]

Additional Paths to Durability: Adaptability vs. Specialization

Human-Centered Design (HCD) can be a highly effective method for understanding human needs to thereby design products that resonate with future customers. Definitive writings on the topic include Henry Dreyfuss' *Designing for People*[21], design firm IDEO's *HCD Toolkit* and the Luma Institute's *Innovating for People*.[22,23] Both offer comprehensive toolsets to usher these processes forward. In an effort to be relevant to paying customers, HCD often provides product developers the rationale to expand their product lines, to meet different user groups' nuanced needs and to justify further product offerings. Sometimes specialization can provide nuanced value to specific user groups like the elderly or the disabled; in other cases, specialization can result in shorter product lifespans when a fleeting or life stage-based need ceases to be relevant.

One area of product specialization is children's products, which include toys, high chairs, utensils, sippy cups and more. High chairs specifically are used while a child is too small and not yet able to sit in an adult chair at the kitchen or dining table. Once made more exclusively out of refurbishable wood or metal, high chairs, like many other contemporary children's products, have become increasingly disposable in their construction. Plastics especially have become a dominant material which can easily stain or discolor. After only a few years of use and by getting soiled by spilled food, they are occasionally safeguarded for years in storage until the next generation. More commonly, they are discarded and end up in the landfill. In the early 2000s, low-cost plastic high chairs with metal tubing were commonly offered by children's product manufacturers. Designs such as these would often be disposed shortly after a child outgrew them. On the other hand, the *Trip Trapp* high chair, produced by Norwegian manufacturer Stokke, was designed to grow with a child so the useful lifespan could be extended. Some owners often continue using them as adult stools for a short period after their children have outgrown them.[24] What if a high chair design could last like a Trip Trapp but graduate to a second more enduring practical use afterward?

One potential strategy is to design products to be *adaptable*. High chairs could oscillate between two functions or return to an original purpose after a families' need for them wanes. In the 1950s, some step stools were designed to also work satisfactorily as high chairs. While children were growing, they could be adapted for mealtime, then folded up or used to reach higher shelves or replace light bulbs. Perhaps there are other specialized products that could be *adaptable*, like these earlier designs, to help transition them to a second life after their initial purpose is no longer required.

Figure 6.4
Above left: An image of a foldable step stool high chair hybrid. Notice the soft cushion on the second step and the metal seat backing to help support a child while eating. Source: Brook Kennedy.

Figure 6.5
Above right: A schematic concept for a high chair step stool hybrid designed to return to step stool utility once a child outgrows the high chair. Source: Brook Kennedy.

Making Disposable Goods Durable Again

For reasons of cost, hygiene, convenience and portability, disposable containers for consumable goods became increasingly popular during the 20th century and continues today. Glass and steel, once the default material for bottling and canning, slowly gave way to aluminum and plastic. The now notorious disposable Polyethylene Terephthalate PET bottle of water, once an icon of on-the-go convenience and perceived safety, is now being more tightly regulated and even banned worldwide as beaches, parks and oceans fill up with them and other plastic waste. Consumers are ever more conscious of this predicament as watchdog organizations have raised awareness of the *Pacific Ocean Gyre*[25] along with scientific reports of micro plastic particles ending up in drinking water and our bodies. To date, science has not yet determined to what extent these microplastics harm human health.[26]

The PET bottle was first introduced to the marketplace in the early 1970s as a lighter and less costly version of glass bottles and cans. Today, more than one million plastic bottles are sold every minute, only a fraction of which are recycled.[27] While some countries have significantly increased plastics recycling, only roughly 30% are recycled in the US, even though a few states and Canadian provinces offer deposits on bottles to encourage their return.[28] Beginning with Oregon in 1972 these bottle bills have helped reduce plastic, aluminum and glass waste but are costly for beverage companies to manage, favoring recycling instead.

Recycling, Upcycling and Downcycling

Designers have looked for ways to keep plastic materials out of the environment by reusing them for furniture and other industrial products. For example, Dave Hakkens, founder of *Precious Plastics*, has developed human-powered, open-source machinery to process and encourage reuse of plastics.[29] To help collect plastic waste dumped in the environment, inventor Boylan Slat's Ocean Cleanup Project developed a highly publicized maritime dragnet used to collect plastic flotsam like fishing equipment. Others have found ways to transform PET into other higher value materials like PET felt which is used in furniture and textile applications.[30] Meanwhile, large multinational food and beverage companies clamor to out-do competitors with incrementally more recycled content in their new PET plastic bottles. More companies are finding ways to promote reusable containers to help end the cycle of consumption, recycling and waste.

Returning to the Reusable Container

Beginning in the 19th century, before PET bottles and contemporary recycling infrastructure, beverage bottles could often be returned to their producers for a deposit. These bottles could then be cleaned and refilled. Soda, carbonated water (seltzer) and milk containers, for example, could be exchanged for a freshly filled replacement as part of a subscription service. Coca Cola long offered their iconic glass bottle in such a format. Certainly, the concept of the milkman delivering glass bottles of fresh milk to one's doorstep and returning to collect them when empty has nostalgic appeal. More to the point, most current life cycle analysis has shown that despite the initial energy and material cost of glass, reusing containers has far less overall environmental impact than effectively recycled plastic.[31] Some smaller dairies have managed to succeed today by reintroducing the reusable glass bottle. With an eye for minimizing processing steps between the dairy cow and the home, distribution costs and impacts have been reduced as well. On-the-go reusable water bottles have also become popular as a response to the perceived impacts of single-use plastic bottles. Dozens of brands now offer stainless steel and reusable plastic water bottles modeled on old fashioned insulated bottles, canteens and flasks.

New business models have also emerged to explore possibilities with packaging reuse. One startup company, Loop (a division of Terracycle), has reintroduced the model of reusable food containers. Whereas food packaging is predominantly disposable or recyclable today with some growing cases of reuse that we discussed previously, Loop is evolving this convention by offering returnable, refillable packaged goods containers made from plastics and metals. When a customer finishes a bottle of water or shampoo that is sold through the Loop platform, they can return the containers to the retailer to be collected, cleaned and refilled. The service is currently available in the US, Canada, the UK, France and Japan, with plans to expand to more countries in 2022. At the time of writing this book, this platform was being piloted with established retailers and restaurants through their eCommerce and brick and mortar channels.[32] Dozens of large consumer packaged goods brands have signed up to deliver their products through this platform.

Other contemporary design-focused startups have found inventive new ways to promote reusability in other categories of consumable goods. Companies producing hand soaps, home cleaning products and personal care items have successfully built businesses based on refillable, reusable containers: Forgo, Blueland, Myro, among many others. In this case, an elegant, durable reusable container can be refilled with a diminutive packet of concentrated soap which when added to water becomes a liquid soap. In the process, material packaging waste is reduced compared with disposable alternatives.

Shipping a small sachet of powdered concentrate also requires far less energy and cost than a fluid soap-filled bottle.

Figure 6.6
Forgo's hand soaps reduce single-use container waste by restoring a refillable, durable-goods business model. The system offers an attractive bottle that consumers desire to save and reuse. Forgo then offers compact sachets of refill soap concentrate that can be mixed with water at home. These sachets can then be shipped inexpensively and ecologically compared with heavier, disposable bottles of premixed liquid soap that are comprised mostly of water. Source: Zea Lindström.

The Rise of Disposable Versions of Durable Goods: The Razor Blade

For those with a passion for cooking, it is unthinkable that one would throw away our favorite kitchen knife once it became dull. Instead, multiple products are available to sharpen knives using traditional techniques. Alternatively, restaurant supply stores and other services are available to sharpen them for us. In many countries in both the developing and industrialized world, foot powered sharpeners, often transported in carts, roam the streets offering sharpening services. On the other hand, many current kitchen gadgets and food preparation products are designed with replaceable blades. This includes the peeler blades of the famous OXO GoodGrips peeler, introduced in 1990. Countless other blades cannot be easily sharpened and are thus theoretically disposable: graters, slicers, corers, etc. In other categories, blades of all kinds can still be sharpened—lawn mower service shops still offer blade sharpening in addition to small motor maintenance.

Face shaving was once performed entirely by hand with a durable, straight razor blade that could be sharpened like a kitchen knife. Traditional barber shops once commonly performed daily shaves, especially when having a clean shaven face was an essential component of professional male hygiene. Barbers would sharpen the blades using a strop such that customers would never feel an unpleasant tug on their skin. Since the invention of the multi-blade razor cartridge in the 1970s, razor blades have become increasingly disposable. Razor blade manufacturers like Schick, Gillette, Harry's and others have been engaged in a race to add more blades to their multi-blade disposable razor cartridges. The decision was born principally out of marketing (as there is little conclusive evidence that a five-blade razor is more effective than a single one). There are also, of course, business reasons for making disposable blades. Making a disposable razor platform requires customers to constantly purchase new ones. The high cost of these disposable blades prompted startups Harry's and Dollar Shave Club to develop "direct to consumer" business models to pass the reduction in retail premiums to the customer (and shareholder). Altogether, the result has been that disposable blades, for which there are limited alternatives, end up in landfills in great quantities. As much as two billion blades are thrown out each year according to a 1990 EPA study.[33] These blades are what *Cradle to Cradle*[34] co-author William McDonough once described as a *monstruous hybrid*, namely, products composed of a few different inseparable materials. In this case disposable razors are composed of plastics, elastomers and stainless steel blades which cannot be separated by a user for recycling.

The original replaceable razor blade, used with "safety razors" (issued to every First World War soldier) could be resharpened many times, even with a knife sharpener, allowing them to last longer than current multi-blade cartridges. In the early 1900s, not long after the invention of the double-edged safety razor, manual and hand-cranked sharpeners were introduced. One model, the Kriss Kross, could sharpen a dull safety razor blade after a simple, careful mounting procedure and a few twists of the crank. Unfortunately, the way in which modern multi-blades are arrayed so closely together make them difficult to sharpen, despite some efforts from inventors.[35] More recently, safety razors have enjoyed renewed popularity, except in most cases they continue to be offered with disposable

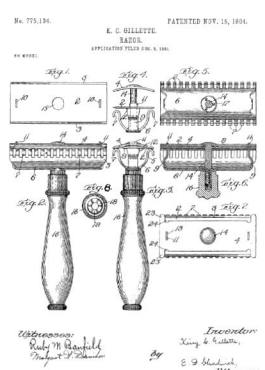

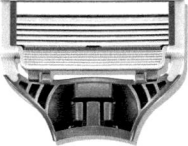

Figuro 6.7
Above left: A patent image of a Gillette safety razor issued to US service men during the First World War. Source: USPTO.

Figure 6.8
Above center: A Kriss Kross hand cranked safety razor blade sharpener. Working like a mechanical strop, these devices could sharpen safety blades almost indefinitely. Source: Joe Haudt (CC BY-SA 2.0).

Figure 6.9
Above right: A typical disposable multi-blade razor cartridge is composed of three or more materials, making it unrecyclable after it wears out after a few weeks of use. Source: Brook Kennedy.

blades. This is an improvement in that a stainless steel blade has the potential to be recycled, but what if the Kriss Kross or a similar design were reintroduced for the current market so that safety razor blades or their modern replacements could be reused many more times, perhaps indefinitely?

Additional Possibility for Reusable Design

There are dozens of other examples of reusable design that have been successfully reintroduced recently. Additionally, there are ample opportunities to further encourage reusability in additional product categories, including coffee shops, food services and others. However, some areas of disposable waste have been more challenging to curtail. Medical waste, specifically, has attracted considerable attention as American hospitals produce five million tons of plastic waste per year or roughly 29 pounds per bed per year.[36] Since the onset of the Covid-19 pandemic, this problem has only deepened with efforts to limit cross infection. During this time, single-use take-away food packaging has boomed as customers avoided on-premises dining. Face masks have also become a new conspicuous form of disposable waste, rivaling the plastic water bottle. Whether speaking of shoes, razors, bottles of soap or beverage containers, returning to durable design from current disposable practices would contribute greatly to reducing landfilled waste—and the excessive consumption of finite natural resources.

Notes

1 Bureau of Economic Analysis (n.d.) *Durable goods*. Available at: www.bea.gov/help/glossary/durable-goods (Accessed: August 10, 2022).

2 The United Nations Conference on Trade and Development (2015) *Circular economy* [online]. Available at: https://unctad.org/topic/trade-and-environment/circular-economy (Accessed: August 10, 2022).

3 Cooper, T. (1994) *Beyond recycling: The longer life option*. London: New Economics Foundation.

4 Waldman, M. (1993) 'A new perspective on planned obsolescence,' *The Quarterly Journal of Economics*, 108(1), pp273–283. https://doi.org/10.2307/2118504

5 Chapman, J. (2005) *Emotionally durable design: Objects, experiences & empathy*. London: Routledge.

6 Pachl, U. and Maurer, S. (n.d.) 'Durable goods. More sustainable products, better consumer rights,' *BEUC*. Available at: www.beuc.eu/durable-goods#whenproductsfail (Accessed: August 10, 2022).

7 Constable, H. (2022) 'Your brand new returns end up in landfill.' *BBCEarth*, January 8 [online]. Available at: www.bbcearth.com/news/your-brand-new-returns-end-up-in-landfill (Accessed: August 10, 2022).

8 Panati, C. (1943) *Extraordinary origins of everyday things*. 1st edn. New York: Perennial Library.

9 Proctor, N. (2018) 'Americans toss 151 million phones a year. What if we could repair them instead?' *wbur*. Available at: www.wbur.org/cognoscenti/2018/12/11/right-to-repair-nathan-proctor (Accessed: August 10, 2022).

10 Matchar, E. (2016) 'The fight for the "right to repair",' *Smithsonian Magazine*. Available at: www.smithsonianmag.com/innovation/fight-right-repair-180959764/ (Accessed: August 10, 2022).

11 European Parliament (2022) *Why is the EU's right to repair legislation important?* Available at: www.europarl.europa.eu/news/en/headlines/society/20220331STO26410/why-is-the-eu-s-right-to-repair-legislation-important (Accessed: August 10, 2022).

12 Morelle, J.D. (2021) *HR 4006 The Fair Repair Act*. Available at: www.congress.gov/bill/117th-congress/house-bill/4006 (Accessed: August 10, 2022).

13 Breslin, N. (2021) *Senate Bill S4104A*. Available at: www.nysenate.gov/legislation/bills/2021/S4104 (Accessed: August 10, 2022).

14 Robinson, B. (2022) 'The cobbling industry is in a dire decline. Jim McFarland has thoughts on how to fix it,' *Stitchdown*. Available at: www.stitchdown.com/stitchdown-shoecast/cobbling-industry-in-decline/?utm_source=rss&utm_medium=rss&utm_campaign=cobbling-industry-in-decline (Accessed: August 10, 2022).

15 Moalem, R.M. and Mosgaard, M.A. (2021) 'A critical review of the role of repair cafés in a sustainable circular transition,' *Sustainability*, 13(22), 12351. https://doi.org/10.3390/ su132212351

16 Park, M. (2017) 'Print to repair: 3d printing and product repair,' in Chapman, J. (ed.) *Routledge handbook of sustainable product design*. London: Routledge. https://doi.org/10.4324/9781315693309

17 Website section on Reuse, focusing on the Riksdagen chair by Åke Axelsson in the early 18980s for the Swedish Parliament. https://garsnas.se/en/reuse/ (Accessed: August 10, 2022).

18 Chapman (2005).

19 Chapman, J. (2014) 'Meaningful stuff: Toward longer lasting products," in Karana, E., Pedgley, O., and Rognoli, V. (eds.) *Materials experience: Fundamentals of materials and design*. Oxford: Butterworth-Heinemann, p142.

20 Chapman, J. (2021) 'Today its cool, tomorrow its junk. We have to act against our throwaway culture,' *The Guardian*. Available at: www.theguardian.com/commentisfree/2021/aug/02/throwaway-culture-products-repair-reuse-recycle-obsolete (Accessed: August 10, 2022).

21 Dreyfuss, H. (2003) *Designing for people*. New York: Allworth.

22 IDEO (2009) *HCD toolkit*. Available at: www.ideo.com/post/design-kit (Accessed: August 10, 2022).

23 LUMA Institute (2012) *Innovating for people: Handbook of human-centered design methods*. Pittsburg, PA: LUMA Institute.

24 The author observed these behaviors qualitatively during the research and development of the award-winning OXO tot Sprout chair which was sold between 2011 and 2022.

25 Lebreton, L., Slat, B., Ferrari, F. et al. (2018) 'Evidence that the Great Pacific Garbage Patch is rapidly accumulating plastic,' *Sci Rep*, 8, 4666. https://doi.org/10.1038/s41598-018-22939-w

26 Vethaak, A. and Legler, J. (2021) 'Microplastics and human health,' *Science*, 371(6530), pp672–674. DOI: 10.1126/science.abe5041

27 Parker, L. (2019) 'How the plastic bottle went from miracle container to hated garbage,' *National Geographic* [online]. Available at: www.nationalgeographic.com/environment/article/plastic-bottles (Accessed: August 10, 2022).

28 US Environmental Protection Agency (2018) *Plastics: Material-specific data*. Available at: www.epa.gov/facts-and-figures-about-materials-waste-and-recycling/plastics-material-specific-data (Accessed: August 10, 2022).

29 Kart, J. (2020) 'This open-source "Precious Plastic" project is changing what waste means and how recycling is done,' *Forbes*, February 12 [online]. Available at: www.forbes.com/sites/jeffkart/2020/02/12/this-open-source-precious-plastic-project-is-changing-what-waste-means-and-how-recycling-is-done/?sh=1db58193f6e8 (Accessed: August 10, 2022).

30 De Vorm's PET felt chairs are made with post-consumer PET. Available at: www.devorm.nl/sustainability (Accessed: August 9, 2022).

31 Coelho, P., Corona, B. and Worrell, E. (2020) 'Reusable vs single-use packaging,' *Reloop Platform & Zero Waste Europe*. Available at: https://zerowasteeurope.eu/wp-content/uploads/2020/12/zwe_reloop_executive-summary_reusable-vs-single-use-packaging_-a-review-of-environmental-impact_en.pdf (Accessed: August 10, 2022). Other categories of containers that can't be recycled or reused are plastic coffee bags with valves. Prior to their recent popularity, coffee was often packaged in recyclable steel cans.

32 Quinn, M. (2021) 'Loop reusable packaging system expands beyond e-commerce to new stores, including Kroger,' *Wastedive* [online]. Available at: www.wastedive.com/news/loop-terracycle-szaky-retail-kroger-mcdonalds-tesco/607272/ (Accessed: August 10, 2022).

33 US Environmental Protection Agency (1990) *The environmental consumer's handbook*. Available at: https://nepis.epa.gov/Exe/ZyPURL.cgi?Dockey=2000URC7.TXT (Accessed: August 10, 2022).

34 McDonough, W. and Braungart, M. (2002) *Cradle to cradle: Remaking the way we make things*. New York: North Point Press.

35 Numerous new gadgets like the "Blade Buddy" promise to sharpen multi-blade cartridges multiple times before being discarded.

36 Kenny C. and Priyadarshini A. (2021) 'Review of current healthcare waste management methods and their effect on global health,' *Healthcare*, 9(3), 284. DOI: 10.3390/healthcare9030284.

Chapter 7

Reimagining Design with Traditional Materials

The industrial revolution, quite simply, transformed the material palette of the built environment. Beginning roughly in the late 18th century, advances in the production of iron and later steel became a staple of tall building construction that continues today. From known landmarks in the 1880s and 1890s like the Eiffel Tower and early tall structures in Chicago and New York through the construction of the Burj Khalifa in Dubai (now the tallest in the world), engineered metals have enabled the construction of monumental structures around the world. Reinforced concrete followed steel in the 1900s. Modernist architects, including those of the Bauhaus, embraced these new materials in addition to glass to transform much of the way commercial buildings are designed today. At present, buildings in the US create 40% of total carbon emissions, 11% of which is produced by the creation of these materials.[1] Construction and Demolition (C&D) waste also accounts for a significant percentage of all material sold waste entering landfills in 2018.[2,3] In automotive design, steel, aluminum and later plastics and a dizzying number of synthetic composites gradually began to replace metals and other material staples. Until the 1950s, cars were made primarily of steel. Today, around 50% of a typical car's volume is plastic, comprising only 10% of its weight.[4] While this weight shift has benefited fuel economy, little of these petroleum-based plastics are recycled. Steel, on the other hand, is more easily scrapped and reused.

The Ford Soybean Car

During the Second World War, the US government passed laws granting the Office of Price Administration (OPA) the ability to ration and reclaim essential resources. A comprehensive scrap collection program was instituted to help gather as much metal, rubber and paper as possible for the production of airplanes, ships, tanks and ammunition to support the war effort. These regulations impacted industrial activity in several ways—while auto manufacturers were enlisted to produce war machinery, these restrictions also impacted civilian business, namely, automobile production. Anticipating the approaching war's impact on the economy, Henry Ford set out to develop an automobile which simultaneously avoided materials that were in high demand for the war effort while also ensuring that such an automobile could operate in a context of gasoline rationing. The result was the Ford "Soybean Car,"[5] which was demonstrated, publicized and later destroyed after being exhibited at the Dearborn Days exhibition. From several anecdotal accounts, the panels of the car's body were made of a resin derived in unknown ratios from soybeans, wheat, hemp, corn and other materials mixed with formaldehyde, although scant records remain of the cars construction to corroborate these claims. Today formaldehyde is considered toxic despite its

DOI: 10.4324/9780367814304-8

durability, yet the potential benefits of vehicles constructed from renewable biomaterials like the Soybean car remain intriguing. Although the material would likely face safety challenges on roads full of heavier steel vehicles, lighter vehicles made of renewable versions of the soybean resin could resist damage from light impact, lower fuel consumption and potentially offer circular economy benefits at the end of a vehicle's life.

Figure 7.1
Ford Soybean concept car c. 1941, Courtesy of the Henry Ford Museum. Henry Ford sought to develop a car that limited use of steel during wartime shortages while advancing efforts to collaborate with agricultural production for manufactured products. Source: The Henry Ford Museum.

Beyond Petroleum-based plastics in Consumer Products

"There are plastics in your toaster, in the blender and the clock, in the lamp and in the roaster, on the door and in the lock, in the washer and the dryer and the garden tools you lend, in your music amplifier and electric fryer—you have got a plastic friend!"

—A promotional rhyme from the 1964 World's Fair[6]

By the late 20th century, petroleum-based plastic had overcome formed and die cast metals and other materials in the construction of smaller consumer products, including accessories, toys, home appliances, children's feeding products, electronics and more. In furniture, plastics and resins began to compete with wood and metal by mid-century, evidenced in the material evolution of the designs of Charles and Ray Eames. Beginning with the development of Celluloid in the 19th century, Bakelite (c. 1907) and eventually petroleum-based polypropylene by the late 1950s, plastics have dominated consumer product manufacture ever since. Because of their utility, cost and flexibility, petroleum-based plastics are now used ubiquitously for disposable goods as well. Unfortunately, today, less than 9% of plastic waste is recycled.[7] What is not recycled has ended up in landfills, oceans, the environment, our food stream and our bodies.[8] While not an indictment of the plastic materials themselves, finding better ways to manage their use are clearly needed. Fortunately, considerable research effort is now aimed at drastically reducing petroleum-based plastic waste including new efforts to promote reusability (as discussed in Chapter 6). In the area of material science, recycled plastics, biodegradable polymers and other classes of materials altogether are now being broadly explored. Many are not new. Some of these materials have been around for ages and are now finding new uses in an effort to mitigate plastic waste. One example, flax, has been showcased in a chair design by Christien Miendertdema. Molded with structural woven flax fibers and plant-based resins, the Flax Chair's materiality is similar to more contemporary glass-fiber composites. Flax fibers are robust and have been used successfully in textiles since the late stone age. They can also be cultivated in many climates around the world.[9,10] Additional biofibers are also being rediscovered for industrial purposes, including hemp and jute, the latter serving as the basis of burlap fabric.[11]

Vulcanized Fiber

Vulcanized fiber, also known as Cottonid in western Europe, is a cellulose-based thermoplastic that was developed in the 19th century, but is relatively unknown in design today. Made from 100% pressed cellulose waste (sometimes paper or cotton waste, hence "Cottonid"), it is hardened into a shock resistant, moldable plastic through a Zinc Chloride parchment process. Effectively purposed for military helmets and luggage in the 19th century, it is now used for safety equipment like welding masks, and for inner layers of laminate furniture.[12] It is also known for its insulating properties in electrical applications. Above all, as an entirely cellulose material, it has a potential end of life strategy—it can biodegrade or be recycled responsibly provided the parchment solution is completely removed. Today, it is still used in consumer electronics to insulate electrical components, like motors and other parts. Since the material can derive entirely from agricultural waste, it shows potential, however the Zinc Chloride solution must be properly contained and managed. Once produced in the US state of Delaware, it continues to be produced in Asia and Europe. Because of its renewability, strength, insulating and moldability potential, its re-adoption in consumer product design is warranted as an alternative to petroleum-based plastics, metals and wood.

Figure 7.2
Consumer products made from vulcanized fiber. While most of these examples are planar, the material can be molded into curved surfaces. Helmets have been constructed with this material in shapes that comfort to a human head. Source: Dagjoh (CC BY-SA 4.0).

Wood Substitutes: Bamboo and Rattan

Forests are harvested globally to provide land for agriculture, livestock and lumber for the building industries. At current rates, forests are declining globally, contributing as well to loss of biodiversity.[13] To help meet demand for timber and related raw materials, Bamboo and similar fast-growing grasses, long used in Asia, are being proposed as contemporary alternatives to conventional lumber. Bamboo is a class of evergreen perennial flowering plant and is the largest species of the grass family. Generally, it is recognized for having numerous benefits as an advanced construction material: it is strong, fast-growing and easy to cultivate in temperate regions around the world on most continents. Bamboo is also a potentially more sustainable alternative to harvesting slower-growing timber forests which have declined overall for decades.[14,15] Critics, on the other hand, worry about the tenacious, invasive qualities of this species as they grow in patches at the expense of other native plants. In Asia and Polynesia, bamboo has been used for thousands of years in architecture and product design and is associated with the visual cultural heritage of these countries. From the framing of traditional houses to the construction of traditional tools, bamboo continues to be widely used. In Hong Kong and China, it is often used as scaffolding in the construction of skyscrapers made of glass, steel, concrete and other contemporary materials.

Figure 7.3
Above left: A traditional Japanese Chasen (Matcha tea whisk) made of bamboo in this manner for more than 600 years. Source: D-Kuru (CC BY-SA 4.0).

Figure 7.4
Above right: The contemporary, computationally designed Bamboo Pavilion by Architects and Researchers Kristof Crolla and Adam Fingrut in Hong Kong. Source: Kristof Crolla.

Only more recently has global interest in the material grown, particularly for flooring and some laminated consumer products for the kitchen. Bamboo researchers Jonas Hauptman, Katie McDonald and Kyle Schulman have argued that bamboo's grain and natural shape carries cultural, aesthetic stigma, leading to poor adoption in the West.[16] Through design explorations, the group has investigated ways to increase its acceptance at the building scale. When combined with contemporary construction techniques, including form-making using digital fabrication and parametric design tools, bamboo use is now maturing and diversifying as evidenced in the contemporary design work of Vo Trong Nghia Architects, LLLab and Kristof Crolla.

Rattan, also a type of grass, has faced similar barriers to global adoption in Western culture, although recently designers have found new ways to increase awareness of the material's renewable benefits. Like Bamboo, Rattan is strong, versatile and fast-growing. In the late 19th through early 20th century, European furniture designers Michael Thonet and Josef Hoffman used Rattan cane to weave durable and pliable seat surfaces. Today, some of these fabrication approaches have become fashionable again, but otherwise Rattan, like Bamboo, suffers from similar aesthetic stigma, often through traditional construction methods like Wicker which carry a strong design period association. In 2017, Jakarta-based industrial designer Abie Abdillah created a line of contemporary Rattan seating for the established Italian furniture company Capellini. Called the Lukis collection, the line combines a contemporary international aesthetic with what the company describes as a "centuries old tradition of craftsmanship."

Figure 7.5
Lukis chair by Indonesian designer Abie Abdillah for Italian manufacturer, Capellini, exemplifies how traditional materials rooted in traditional production can be successfully re-envisioned through design to compete with synthetic materials for wider appeal. Source: Cappellini, Abie Abdillah, Designer.

Renaissance in Glass after Bisphenol A

"If we are going to live so intimately with these chemicals eating and drinking them, taking them into the very marrow of our bones – we had better know something about their nature and their power."[17]

—Rachel Carson, speaking about pesticides in *Silent Spring*

Glass has evolved over thousands of years for decorative, utilitarian and architectural use (see Chapter 4). In food packaging, storage and transport, it continues to be used for jars and bottles and is, of course, commonly used for drinking glasses and other table ware. Clear plastics, like polyethylene (LDPE and HDPE), polypropylene (PP) and polycarbonate (PC), have replaced glass in many of these areas for reasons of cost, weight and user-convenience. Unlike glass, plastic containers don't shatter when dropped. As a result, these plastics were often selected for children's feeding products as toddlers learn to wield adult drinking glasses.

Beginning as early as the 1930s, studies suggested that Bisphenol A (BPA), a substance found in some plastics like PC and used in linings for canned food, had some level of human toxicity. This discovery would begin a decades-long tug of war between industry, government and health officials trying to balance the utility of these plastic materials with justifiable health concerns. Despite mounting evidence connecting BPA with health issues—notably endocrine disruption, child brain development and prostate problems[18]—products that contain the chemical remain unregulated and on the market. In 1999, the US-based nonprofit Consumer Reports found BPA in liquids in baby bottles that were heated. In the beginning of 2007, congress launched an investigation of government conflict of interest in policies about BPA safety. The result of these proceedings led to public health concerns about PC specifically, and a wave of safety labeling aimed at reassuring customers that plastic baby bottles and water bottles were "BPA Free." Because of the conflicting accounts of the material, many manufacturers ceased to use PC, including *Nalgene*[tm] who had popularized the material through their indestructible outdoor sports and camping water bottles. As of writing this book, many countries have banned BPA outright, whereas others have for specific applications, like baby bottles. Other measures have sought to limit its use in canned food packaging but not ban it outright.

As justifiable concerns about products made with BPA-leaching chemicals have lingered, entrepreneurs and existing manufacturers turned to plastic alternatives. Stainless steel bottles, for example, have experienced a resurgence even though it was revealed that some manufacturers had BPA plastic interior linings, like canned food. Simultaneously, glass has experienced a comeback, especially for making warm beverages. As a material that has existed for thousands of years, no known health problems were associated with its use (with the exception perhaps of leaded glass).

Cookware and Convenience: The Rise and Fall of Non-Stick Surfaces

Cleaning pots and pans after cooking has long been considered a time-consuming and undesirable chore. As with many domestic tasks, inventions and technologies of all kinds were introduced in the mid 20th century promising to make chores easier. The promise of effort-saving kitchen innovation persists. Scrubbing tools, using wire and nylon brushes, steel wool and nylon mesh, have all become profitable businesses for addressing these user troubles. However, non-stick cookware might be among the most successful technologies ever created for the kitchen. In 1938, Dr. Roy Plunkett accidentally discovered PTFE while developing commercial refrigerants for the DuPont company. Inert to most chemicals, PTFE is among the most slippery substances known to mankind. By the end of the Second World War, PTFE was rebranded *Teflon*[tm]. The substance was also used in many other industrial applications and continues to be used today. Pots and skillets made with the technology enabled cooking with minimal cleanup, as no food, no matter how burned (or overcooked), would stick to the surface.

Despite the user benefits of Teflon-coated pans, several concerns emerged about the safety of chemicals used to produce the material. In particular, Perfluorooctanoic acid (PFOA) has concerned health officials and consumers alike because it does not break down and has ended up in most Americans' blood streams.[19] In some areas of the US, PFOAs and their larger chemical category PFAS were discarded in the environment resulting in elevated public exposure to the chemical. In high concentration, a panel of epidemiologists found a connection between PFOAs and numerous serious health ailments, including cancer, liver problems and decreased fertility, among others.[20] While minimal traces of PFOAs have been found on non-stick cookware, consumer opinion has partially turned against non-stick coatings, spurring a renaissance in more traditional cookware materials: cast iron, stainless steel, enameled cast iron and ceramics have all seen a boost in consumer demand as of the writing of this book. Several startup companies, including Caraway, offer an "innovation," back to enamels long used and tested in kitchens. Similar enameled cast iron designs including those of French manufacturer Le Creuset began to be produced in 1925. Originally, enameling metals for cookware dates to the mid 18th century when chemists and industrialists began finding ways to bond enamel to metals. Some of the first enameled cookware and stoves date to the 18th century. While likely not as effective as Teflon and related material coatings, many of these earlier and revived cookware technologies have capable "non-stick" qualities while protecting (and reassuring) customers with material safety.

Monstruous Hybrids: The Rise of Fused Multi-Material Consumer Products

As this book discussed earlier in Chapter 6, there is tremendous value in designing consumer goods to be repaired, refurbished and maintained. However, in many categories of durable consumer goods, often in the kitchen, food preparation tools, from spatulas to vegetable peelers inevitably wear out. Perhaps a blade becomes dull, or the material becomes soiled or discolored from sustained use. Some can be repaired. Many brands of peelers offer replacement blades to perpetuate their usefulness. However, once a utensil is no longer useful, many modern designs are disposed because their multi-material assemblies cannot be separated into individual materials for recycling. Many of their component materials cannot be recycled anyway. These "Monstruous Hybrids," as William McDonough and Michael Braungart define them in their book *Cradle to Cradle*, are destined for the landfill.[21]

Celebrated industrial design firm Smart Design, known as a pioneer and leader in *inclusive design*, has been widely influential in the design of product handles of all kinds since the early 1990s.[22] Notably, they designed the popular OXO GoodGrips vegetable peeler (and many other products for their successful consumer product brands). Principal to the design's success is the ergonomic handle, recognizable for its generous size and rubbery elastomeric material. Often GoodGrips handles are comprised of two parts: a Polypropylene or Nylon plastic core on top of which the elastomeric material is molded. This construction has influenced the handle design of all kinds of products, from handheld electronics to medical instruments. Designed originally for users with special needs, specifically those with limited hand dexterity from arthritis, the GoodGrips handle has been extensively copied for its inclusive philosophy and gratifying, tactile feel. As a result, the GoodGrips peeler is hailed as an Industrial Design milestone in permanent collections of numerous museums, including the Museum of Modern Art (MOMA) in New York.[23]

Indeed, this approach to design is admirable for helping those with disabilities, but the execution of handle designs with multi materials also comes at an environmental cost. In the case of OXO's products, they are designed to last a lifetime. On the other hand, so many disposable products have ended up with handles like these (toothbrushes come to mind). Even with the most long-lasting products, being recyclable is beneficial in the event the product is damaged, broken, soiled or no longer valued. Moreover, there are many applications where a secure over-molded grip is not always necessary. Peelers require a secure hand grip to operate, whereas other tools, like soup ladles for instance, where the grip interaction is more relaxed, might not need the same kind of handle design. Of course, product managers unify design features across product lines to promote consistent visual brand language in the marketplace, regardless of whether their utility is duly essential.

Prior to the development of the multi-material handle, tools of all kinds employed simpler handle assemblies, some, of course, using single materials like metal and wood. Today, this kind of traditional construction has reemerged. OXO's own line of wood kitchen utensils are an excellent example. All-metal tools have also become widely used again for their professionalism and perceived material safety. Today, tools made entirely of stainless steel or aluminum can be more easily scrapped for reuse or recycling. Now that cast iron, stainless steel and enamel cookware have also seen a resurgence again, there is less concern about scratching non-stick cooking surfaces. Can handles provide a good ergonomic grip without sacrificing a sound end-of-life strategy? Some manufacturers have sought to create this balance: Kuhn Rikon's version of the now famous "Y" peeler, for instance, creates a secure grip using a geometric indentation rather than relying on a soft, elastomeric over-molded handle.

Notes

1 Zhong, X., Hu, M., Deetman, S. et al. (2021) 'Global greenhouse gas emissions from residential and commercial building materials and mitigation strategies to 2060,' *Nature Communications*, 12, 6126. https://doi.org/10.1038/s41467-021-26212-z

2 Bossink, B.A.G. and Brouwers, H.J.H. (1996) 'Construction waste: Quantification and source evaluation,' *Journal of Construction Engineering and Management*, 122(1), pp55–60.

3 Environmental Protection Agency (2020) 'Advancing sustainable materials management: 2018 fact sheet.' Available at: www.epa.gov/sites/default/files/2021-01/documents/2018_ff_fact_sheet_dec_2020_fnl_508.pdf

4 Cumbers, J. (2020) 'Your car is more plastic than you know. This company may soon be brewing its parts with biology,' *Forbes*, August 1 [online]. Available at: www.forbes.com/sites/johncumbers/2020/08/01/can-this-california-startup-brew-sustainable-guilt-free-plastic-from-sugar/?sh=29237f767070 (Accessed: February 1, 2022).

5 *The New York Times* (1941) 'Ford shows auto built of plastic; strong material derived from soybeans, wheat, corn is used for body and fenders. Saving of steel is cited. Car is 1,000 pounds lighter than metal ones,' *The New York Times*, August 14 [online]. Available at: www.nytimes.com/1941/08/14/archives/ford-shows-auto-built-of-plastic-strong-material-derived-from-soy.html (Accessed: August 11, 2022).

6 *Broadway Classics* (n.d.) 'The wonderful world of chemistry – Dupont – 1964 New York World's Fair,' *YouTube*. Available at: www.youtube.com/watch?v=wQCKqDJksuE (Accessed: August 11, 2022).

7 Loria, K. (2021) 'The big problem with plastic: CR reveals where most of the plastic you throw away really ends up and explains what to do to limit its environmental harm,' *Consumer Reports*, September 8 [online]. Available at: www.consumerreports.org/environment-sustainability/the-big-problem-with-plastic/ (Accessed: August 11, 2022).

8 Parker, L. (2022) 'Microplastics are in our bodies. How much do they harm us?' *National Geographic*, April 25 [online]. Available at: www.nationalgeographic.com/environment/article/microplastics-are-in-our-bodies-how-much-do-they-harm-us (Accessed: August 11, 2022).

9 Solanki, S. (2018) *Why materials matter: Responsible design for a better world*. London: Prestel Publishing.

10 Franklin, K. and Till, C. (2019) *Radical matter: Rethinking materials for a sustainable future*. London: Thames & Hudson.

11 Murali, B., Yogesh, P., Karthickeyan, N.K., and Chandramohan, D. (2022) 'Multi-potency of bast fibers (flax, hemp and jute) as composite materials and their mechanical properties: A review,' *Materials Today: Proceedings*, 62(4), pp1839–1843.

12 Scholz, R., Delp, A. and Walther, F. (2020) 'In situ characterization of damage development in cottonid due to quasi-static tensile loading,' *Materials*, 13(9), 2180. Available at: https://doi.org/10.3390/ma13092180 (Accessed: August 11, 2022).

13 Nunez, C. (2016) 'Deforestation facts, deforestation information, effects of deforestation – National Geographic,' *National Geographic*, March 30. Available at: http://environment.nationalgeographic.com/environment/global-warming/deforestation-overview/ (Accessed: August 11, 2022).

14 Food and Agriculture Organization of the United Nations (2020) 'State of the world's forests.' Available at: www.fao.org/state-of-forests/en/ (Accessed: August 11, 2022).

15 Pearce, F. (2018) 'Conflicting data: How fast is the world losing its forests?' Available at: https://e360.yale.edu/features/conflicting-data-how-fast-is-the-worlds-losing-its-forests (Accessed: August 11, 2022).

16 Schumann, K., Hauptman, J. and MacDonald, K. (2019) 'Addressing barriers for bamboo: Techniques for altering cultural perception,' in *ARCC Conference Repository*, 1(1). Available at: www.arcc-journal.org/index.php/repository/article/view/664

17 Carson, R. (1962) *Silent spring*. Boston: Houghton Mifflin.

18 Fenichel, P., Chevalier, N. and Brucker-Davis, F. (2013) 'Bisphenol A: An endocrine and metabolic disruptor,' *Annales d'Endocrinologie*, 74(3), pp211–220.

19 Agency for Toxic Substances and Disease Registry, Center for Disease Control (2020) 'PFAS in the US population.' Available at: www.atsdr.cdc.gov/pfas/health-effects/us-population.html (Accessed: August 11, 2022).

20 US Environmental Protection Agency (2022) 'Our current understanding of the human health and environmental risks of PFAS' [online]. Available at: www.epa.gov/pfas/our-current-understanding-human-health-and-environmental-risks-pfas (Accessed: August 11, 2022).

21 McDonough, W. and Braungart, M. (2002) *Cradle to cradle: Remaking the way we make things*. New York: North Point Press.

22 Shelley, K. (2011) 'Simply smart design,' *Cooper Hewitt* [online]. Available at: www.cooperhewitt.org/2011/02/04/simply-smart-design/ (Accessed: August 11, 2022).

23 The author was employed by Smart Design and by extention OXO from 2005 2012 where he designed dozens of products under this partnership.

Chapter 8

Manual and Therefore Modern?

Smart products, sometimes referred to as the *Internet of Things* (IOT), have flooded the marketplace in the 2010s following the successful introduction of smartphones a few years earlier. Combined with smartphone-based software control, this category of consumer products leverages sensors, wireless communication and micro-processors to control a dizzying amount of home automation tasks: thermostats (notably Nest thermostats and competitors), front door cameras, baby monitors, health monitors, air quality monitors, adaptive smart shades and more. Google and Amazon produce devices to control them all with voice activation. Without question, many of these devices have helped and brought conveniences to users. Not to be left behind, traditional makers of durable goods have joined the IOT craze and vied, in turn, to integrate smart control into existing "dumb" appliances such as refrigerators, washing machines and cars. Some auto makers have even abandoned the analog metal door key for evermore wireless control, with mixed success.[1] Other advanced sensing systems have greatly improved car safety through air bags, dynamic stability control and through services like General Motor's OnStar which connects motorists to emergency first responders.

In the book *Trillions: Thriving in the Emerging Information Ecology*, authors Peter Lucas, Joe Ballay and Mickey McManus ponder the opportunities and impact of a world fully embedded with internet-connected sensors and micro-processors.[2] Others, including government regulators, have been more cautious. It is well known how some technology companies have used IOT devices, phones and desktop computers to collect valuable data about people's behavior.[3] Beyond these concerns, allowing so many parts of our lives to be controlled and powered by electronics, smart or otherwise, comes with some environmental tradeoffs. In addition to questions of obsolescence cycles, durability and repairability (as discussed in Chapter 7), smart and electrified appliances, in aggregate, also contribute significantly to per-capita energy use at home. This does not include the additional energy used in the background through data server infrastructure that enables the internet and cloud to run in perpetuity. Speaking about in-home energy use, a 2018 study led by the Environmental Protection Agency found that US home energy consumption steadily increased between 1980–2015 all from the growth in *miscellaneous* electrical appliances.[4] This includes televisions, computers, mobile phones, coffee makers and IOT systems. Some of these products plug in, others rely on rechargeable batteries.

DOI: 10.4324/9780367814304-9

Rise of the Rechargeable Battery

Rechargeable batteries have slowly proliferated in home goods in the past three decades, notably in mobile phone technology. Around the globe, sales of electric vehicles (EVs) have reached significant numbers after several prior attempts to enter the market over the past 130 years. Improved batteries have made this possible, and as a consequence, demand for EVs has soared. In many respects, this is good news. Electrification of cars, trucks and other vehicles promises to cut tailpipe emissions significantly when powered by renewably generated electricity.[5] At the same time, the batteries that enable them to work remain an environmental quandary: disposable and rechargeable batteries alike have piled up in landfills where they have leached toxic chemicals into groundwater.[6] As lithium-ion batteries have advanced considerably in their storage capacity, their material sourcing, production and disposal have created significant environmental and social problems.[7] Addressing these concerns have become top priorities for industries and government regulators alike and will likely lead to new ways to solve these shortcomings. In the meantime, batteries, especially lithium-ion, are finding their way into every part of the home.

In the 1860s, English economist William Jevons observed how advancements in using coal more efficiently led, non-intuitively, to *increased* consumption. Instead of these industrial efficiency gains leading to a decrease in the amount of coal used, the very opposite happened. This concept is now known as the *Jevons Paradox*.[8] Similarly, as batteries and electricity use become more efficient, consumption too appears to be increasing. Considering the phenomenon of the Jevons Paradox, do all products and appliances really need to be electrified? Can some of the consumer products in our lives remain at least partially human-powered or perhaps return to pre-electric propulsion without losing human acceptance? More importantly, can design make hand-powered alternatives acceptable if not *more desirable* than their electric alternatives? This chapter will explore this question further.

Human-Powered Electronics: Cameras, Razors, Phones and More

Many tasks performed by smaller electric devices, including some IOT products, were once accomplished with manual human power, just as the push lawn mower and manual carpet sweeper predated their gas and electric-powered replacements. Certainly, many of the benefits of some smaller technological and IOT devices cannot be easily manualized. A Nest thermostat, for instance, might not be easily replaced by a hand-powered alternative as it draws a small amount of electricity 24 hours a day. On the other hand, many devices that use battery power only while in use could potentially be supplemented (or perhaps replaced) with some form of manual override, allowing users the option to power them with their own muscle power. In the process, such an approach could reduce miscellaneous power consumption and provide some light physical activity. Here are some potential examples:

Powered/Battery-Powered Device	Manual Precedent
Small Yard Gas or Electric Lawn Mower	Push Lawn Mower
Stick Vacuum Cleaners	Carpet Sweepers and Brooms

Blenders	Manual Blenders
Automatic shavers	Wind-up shavers
Electric toothbrushes	Wind-up toothbrushes
Electric towel dispensers	Pull out towel dispensers

Wind-Up Appliances

Film cameras, the analog precursor to the digital video camera and smartphone, were originally hand-wound before mobile electrical power was practical with batteries. In the 1920s, cameras became small enough to take hand-held ergonomics into consideration. The Swiss-made Bolex camera, for instance, was introduced using half-width 16mm film. These kinds of smaller cameras addressed the demand from both the growing newsreel and documentary fields, as well as the emerging amateur market. They were specifically designed to use shorter lengths of film, measuring 100 to 200 feet, and could be hand-wound using mainspring clockworks. One winding could power the camera continuously through most of a film roll. These cameras saw some use in early professional filmmaking.[9] Obviously, film cameras are no longer used—nevertheless, what other small appliance could be powered with wind-up mechanisms like these or at least have that option if a battery were to run out?

Wind-Up Automatic Shavers

The first functional electric shavers were patented near the turn of the 20th century by German and American inventors. Later, Jacob Schick, of the eponymous Schick razor brand that survives today, patented a similar device in 1930. Shortly thereafter, the plug-in rotary electric shaver was patented by Philips N.V. in the Netherlands from the invention of Alexandre Horovitz and Alexis Van Dam.[10] By the 1960s, electric razors became widespread. Convenience likely led to their popularity—they don't require water, a sink, shaving cream and expensive disposable blades that wear out quickly. Yet for their small size, electric shavers still require a plug, and they use electricity. Similarly, battery-powered shavers require disposable or rechargeable batteries. Some contemporary electric shavers even have internal batteries that can't be replaced or removed before their disposal. Electric shavers with failed internal batteries are summarily thrown out. Rechargeable electric toothbrushes and other small electronics also suffer from similar shortcomings as discussed in Chapter 6.

In the 1950s and 1960s, a few manufacturers introduced wind-up automatic shavers as an alternative. These used no batteries and could shave automatically for a few minutes by wind-up power alone. Popular in the former republics of the Soviet Union and Europe, US-based Haverhill distributed a model that was manufactured in Monaco and later traveled aboard the Apollo 14 Lunar space mission in 1971.[11] Forgotten for decades, the author of this book tested one in a class environment to ascertain its practicality. With encouraging results, the author then offered a former industrial design student David Shaltanis a brief to develop a "contemporary" version of a wind-up razor for his senior graduation project. Called the *Moove* razor, Shaltanis explored how design could make wind-up technology "modern" and appealing to users again. Taking some timeless cues from Dieter Ram's midcentury electric razor designs for Braun, the *Moove* razor aspires to reimagine this alternative technology for 21st-century design. The result was a critical success, receiving many honors and awards, including a 2021 IDEA Gold Award from the Industrial Design Society of America.

Figure 8.1
Above left: A Haverhill Shaver like the one that was used by astronauts Alan Shepard Jr. and Stuart Roosa in 1971 with tremendous promotional fanfare following the Apollo lunar mission's success. Source: From page 64 of the November 26, 1971 issue of Life. Scan by Google Books. Extraction and clean-up by heroicrelics.

Figure 8.2
Above center: Designer David Shaltanis' award-winning *Moove* razor, a contemporary wind-up design concept. Source: David Shaltanis.

Figure 8.3
Above right: The *Moove* razor's form factor follows the spiral mainspring clockwork which stores the potential mechanical energy to run the cutting blade. Source: David Shaltanis.

Manual Towel Dispensers

As one of many measures to avoid contracting the flu or more serious ailments, the US Center for Disease Control (CDC) recommends washing your hands for 20 seconds after using the restroom. Drying your hands, though, raises additional questions: some evidence suggests that towels are more effective than hand driers, but the data are not unanimous.[12] For many paper towel dispensers, you have to turn a crank or press a tab with your hand to release a towel, potentially compromising the sanitary reasons for washing your hands in the first place. Some will try awkwardly to use their elbows to avoid touching the dispenser. Touch-free electric dispensers offer a solution but are generally battery powered. Sometimes they either run out of battery power, jam or break, rendering them useless. In contrast, older folded towel dispensers manage to pre-load a fresh towel after one is retrieved, without cranks or electricity, by making use of the interwoven folds in their towel's stacked design. When an electrically dependent appliance like a towel-roll dispenser runs out of charge, a manual override would be useful. Could these automatic dispensers not be designed with an intuitive, hands-free lever so a towel could be retrieved no matter what?[13]

Manual Override: Hybrid Hand Drills, E-Bikes and More

Today, hand-held power drills with rechargeable batteries or plugs are an essential tool in the building industries across furniture-making, hobbyist and construction markets. As batteries get smaller, lighter and more powerful, they will surely be able to replace corded drills entirely. At the same time, the act of drilling holes far predates the electric motor and battery. Specifically, in the 19th century, hand-cranked drills developed rapidly, built from metals and wood.[14] For softer materials like wood, these tools performed reliably, even if they might wear out the user from repetitive use. By contrast, electric drills powered exclusively with batteries become useless when their batteries run out, including back-up batteries. Perhaps electric drills and other electric tools could be redesigned with some form of manual override. Similar in principle to hybrid-electric bicycles that can be pedaled when the battery discharges, manual override power could be useful in pinch. This could also extend the usefulness of battery-powered tools beyond the point at which compatible replacement batteries are no longer available.

Figure 8.4
An E-Bike, capable of being pedaled, used under electric propulsion or a combination. Source: bikesharedude.

Manual Override: The Evolution of the Murphy bed to the "Murphy Room"

Murphy beds, used in auxiliary rooms in apartments or single room studios, enable small apartments to maximize their limited square footage by providing convertible space (living space vs. sleeping space). In practical terms, a Murphy bed folds down from a vertical position next to a wall or within a built-in cabinet, to a horizontal position so people can sleep. First invented in the early 20th century by W.L. Murphy[15] in San Francisco, the original purpose of the Murphy bed was to circumvent religious taboos about bringing romantic partners into a "bedroom"—a condition which could not be technically achieved in one room apartments. Generally, Murphy beds disappeared, especially in the US, as homes grew larger and people moved to the suburbs for more space. Now, as American cities have grown popular again, living in small quarters and an interest in maximizing space in urban environments has returned. Startup company Ori Systems has pushed the concept of the Murphy bed into what might be considered a "murphy room," essentially a moving desk, shelf and bed cabinet that can transform a space from a sleeping room to a living room or den. Ori describes itself as a "robotics company." Unlike a Murphy bed that relies on human effort to deploy, Ori is fully automatic. Voice, phone and remote control activate the system electrically. In addition, it relies on sensors to prevent collisions with objects in the way. Indeed, the Ori system provides benefits for the small home owner to maximize space. Does the system's electrification make it vulnerable to failure and force needless energy use and frustration?

Both the *Studio* and *Pocket* collections slide on the floor, in a way similar to space-saving mobile shelving which is used in libraries and archives worldwide. With mobile shelving technology, library floors can save up to 50% of floor space, using hand cranks to move the heavy book stacks manually without needing sensors, motors or electricity. Additionally, the modest physical motion provides some healthy activity which is neither strenuous nor dangerous. Perhaps a similar model could be made of an Ori system offering a *manual override*, enabling a user to use their muscles if possible or when desired, while reducing energy consumption in the process.[16]

Figure 8.5
Above left: An Ori Systems "Pocket Studio" with user activated robotic control to open and close from bedroom to living room mode (pictured). Source: Ori Design Studio, www.oriliving.com.

Figure 8.6
Above right: A mobile shelving roller unit that moves in a similar way but without motorization. Source: Newcastle Libraries.

Battery-Powered Bicycle Accessories vs. Manual Alternatives

Dynamo-powered bicycle lamps were developed in the 19th century not very long after bicycles became a popular form of mobility. Decades later, dynamo lights remain common in European cities that define bicycles as registered vehicles alongside cars and trucks. As suburban developments expanded in the US and other countries, commuter bicycling declined next to automobile use and other forms of transportation. Bicycle lighting developed into an optional form of safety, particularly for recreation-focused cycling. These lights use batteries, and eventually rechargeable batteries with LEDs. In the meantime, European bicycles are often equipped with dynamo lights from the factory to comply with stricter safety regulations. Some even have dynamos integrated into the wheel hubs. Slightly bigger in size, they lack the battery recycling challenges previously discussed. They are also a more reliable safety feature as there is no battery to wear out or discharge. Many of these systems can therefore last for decades without maintenance, except for bulb replacement. When the bicycle is scrapped, the housing and dynamo, made mostly of metal materials, can be more easily reused and recycled (with the exception of the bulbs). Manual bicycle bells, like dynamo-powered lights, also rely completely on human power to operate. This is extremely important for safety. San Francisco-based Spurcycle introduced a modern finger-powered bell that produces a clear, pleasant yet penetrating sound that can be easily heard. Manufactured in part by a company that has produced bells since 1854, there are no batteries that can fail or be discharged when you need them the most. They are also produced durably to last for decades.

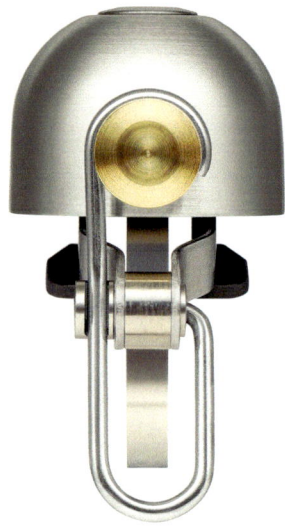

Figure 8.7
A Spurcycle bicycle bell. Using a bell housing designed in 2013 but made in a Connecticut facility specializing in bell manufacture since the mid 19th century. The ring tone is loud, clear and pleasant without using electricity. Source: Spurcycle.

Thinking Manually

Examples in this chapter describe products that rely at least to some extent on human power, building on related examples already shared in Chapter 5 (push lawn mowers, carpet sweepers, etc.). Some of these examples were introduced before electronics, motors and electricity were available as they are today. Other examples were designed to be appropriate for the conditions of unique environments, including space travel and contexts that lack grid electricity. Undoubtedly there are further examples of the potential of manual power that have not been covered here. Altogether, they suggest additional ways to conserve energy and propose future design possibility. On the other hand, it is perhaps true that not all manual design would be accepted by users today, especially those with disabilities and others who are looking for comfort and convenience. However, emerging contemporary science is now suggesting that relentless comfort and convenience, born of automatic electrification, has its limitations as well, especially in terms of health. For instance, Dan Buettner's influential study of *Blue Zones*, about communities around the world with the longest life expectancy, offers a sobering picture: these communities tend to integrate light daily physical activity into their daily routines (often walking and light chores).[17] Additional studies have found evidence that light physical activity can also be beneficial to cognitive function.[18] Perhaps design could help encourage a better balance between light physical activity and electrified convenience through the manual technologies discussed here—to improve health while simultaneously conserving energy. Additional benefits are apparent as well: when the power goes out or batteries run out, they can still be used reliably.

Notes

1 Blanco, S. (2019) 'Tesla owners locked out of cars on labor day when phone key app goes down,' *Car and Driver*. Available at: www.caranddriver.com/news/a28904319/tesla-owners-locked-out-of-cars-phone-key/

2 Lucas, P., Ballay, J. and McManus, M. (2012) *Trillions: Thriving in the emerging information ecology*. New York: Wiley.

3 Bensinger, G. (2021) 'Google's privacy backpedal shows why it's so hard not to be evil,' *The New York Times*, June 14. Available at: www.nytimes.com/2021/06/14/opinion/google-privacy-big-tech.html (Accessed December 9, 2022).

4 Fanara, A., Clark, R., Duff, R. and Polad, M. (2007) 'How small devices are having a big impact on U.S. utility bills,' *Energy Star*. Available at: www.energystar.gov/ia/partners/prod_development/downloads/EEDAL-145.pdf (Accessed: January 25, 2022).

5 U.S. Department of Energy (n.d.) *Emissions from electric vehicles*. Available at: https://afdc.energy.gov/vehicles/electric_emissions.html (Accessed: April 21, 2022).

6 Jacoby, M. (2019) 'It's time to get serious about recycling lithium-ion batteries,' *Chemical and Engineering News*, 97(28) [online]. Available at: https://cen.acs.org/materials/energy-storage/time-serious-recycling-lithium/97/i28 (Accessed: April 21, 2022).

7 Editorial (2021) 'Lithium-ion batteries need to be greener and ethical,' *Nature*, 595, p7.

8 Polimeni, J.M. (2008) *The Jevons Paradox and the myth of resource efficiency improvements*. London: Earthscan.

9 Salt, B. (1992) *Film style and technology: History and analysis*. London: Starword.

10 Horowitz, A. and Van Dam, A. (1940) 'Haarschergeraet,' *Espacenet*. Available at: https://worldwide.espacenet.com/patent/search/family/025761830/publication/DE694507C?q=pn%3DDE694507 (Accessed: August 11, 2022).

11 Manual hair clippers using scissor-style mechanisms also predated electric models which are now a staple tool in barber shops.

12 Huang, C., Ma. W. and Stack, S. (2012) 'The hygienic efficacy of different hand-drying methods. A review of the evidence,' *Mayo Clin Proc*, 87(8), pp791–798. DOI: 10.1016/j.mayocp.2012.02.019. Additional studies have demonstrated that despite perceptions in the US, cleanable, reusable scrolling cloth rolls can also be effective and sanitary. While they are used in Europe they have almost entirely disappeared in the US.

13 After the introduction of electric starter motors on motorcycles, kick starters were removed on many models, which were useful when batteries were too low to start the engine.

14 Sloane, E. (1965) *A museum of early American tools*. Mineola: Dover Publications.

15 Murphy, W.L. (1912) *Disappearing bed*. Available at: https://patents.google.com/patent/US1000201 (Accessed: August 11, 2022).
16 Rear hatches on cars are often automatic and cannot be operated manually without damaging the automatic system.
17 Buettner, D. (2008) *The blue zones: Lessons for living longer from the people who've lived the longest*. Washington, D.C.: National Geographic.
18 Lee, S.Y., Pang, B.W.J., Lau, L.K., et al. (2021) 'Cross-sectional associations of housework with cognitive, physical and sensorimotor functions in younger and older community-dwelling adults: the Yishun Study,' *BMJ Open*, 11: e052557. doi: 10.1136/bmjopen-2021-052557

Chapter 9

Manufacturing with Water, Wind and Human Power

Before industrial manufacturing relied on coal and other fossil fuels, labor-intensive tasks were performed using power from water, wind, animals and human beings.[1] From grist mills powered by flowing river water to the earliest windmills of Asia, Persia and Europe, the forces of nature were once entrusted to grind flour, cut wood, pump water and manufacture industrial products.[2] Some evidence suggests that wind-powered mills date back thousands of years to Egypt, China and the Roman Empire, although not always for agricultural purposes.[3,4] Between the 7th and 9th centuries AD, the "horizontal" windmill appeared in Persia, a design configuration with a vertical axle and vanes placed around it in a radial pattern. Later, the technology emerged in Europe in the 12th century, possibly brought from Asia after the Crusades.[5] Between the later part of the 13th century and the mid 19th century, the tower windmill emerged, with the familiar multi-blade propeller

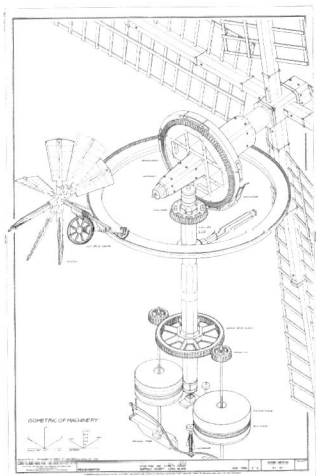
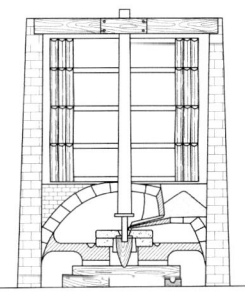
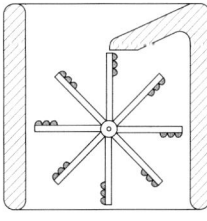

Figure 9.1
Above left: The Beebe Windmill, Bridgehampton, New York, 1820. A relatively recent example of a 19th-century windmill encapsulating hundreds of years of Persian and European improvements to the technology. Source: Kathleen S. Hoeft and Chalmers G. Long Jr. 1976 HAER, National Park Service.

Figure 9.2
Above right: A schematic diagram of a Persian horizontal windmill. This type of windmill precedes the more familiar configuration used in contemporary wind-turbines. Source: Kaboldy (CC BY-SA 3.0).

DOI: 10.4324/9780367814304-10

rotating with the wind perpendicular to the ground. During this time, this basic type of windmill became tremendously popular in Europe and the technology slowly became more powerful and efficient. In the 15–17th centuries, Dutch artists from Jan Van Eyck to Rembrandt Van Rijn portrayed windmills as a symbol of the country's culture and economic productivity.[6] At the technology's peak, as many as 200,000 windmills operated across the European continent, especially in areas where rivers could not be harnessed for mechanical power.[7] By the 19th century, with the arrival of the steam engine, the number of windmills quickly declined. By the early 20th century in the Netherlands, a fraction of these windmills remained of more than 10,000 that had operated in their prime. Many that survive today are preserved as museums or are inoperable. In contrast, the 9th-century horizontal windmill continues to be used in Iran for the same purposes it had served hundreds of years earlier.

From Windmills to Windpumps

European colonists brought windmill technology to North America. Following the Louisiana Purchase in 1803, the US expanded westward, claiming additional ancestral Native American territory. Much of these lands were settled for farming from the great plains to California. Windmill technology was adapted to pump water from the ground for agriculture, consumption and sanitation. In 1854, machinist Daniel Halladay invented a self-regulating windpump

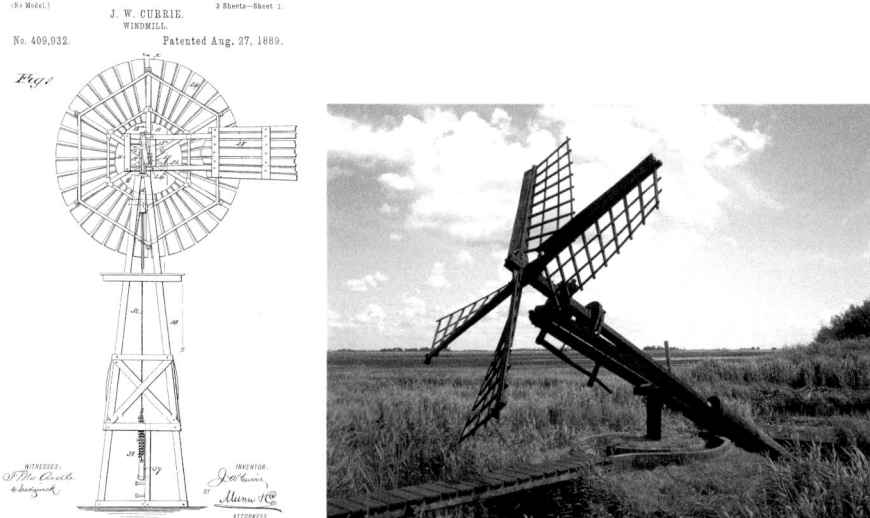

Figure 9.3
Above left: An 1889 US patent for a windmill. Windmills and windpumps were used to pump well water to the surface for agricultural irrigation continuing hundreds of years of Persian and European improvements to the technology. Source: J.W. Currie.

Figure 9.4
Above right: A Dutch wind-powered Archimedes pump (Tjasker) used for ages to drain lowlands. Such an invention might be inspiring for residents and government agencies in coastal communities of the Gulf of Mexico where routine flooding and power outages are common after hurricanes. Source: TUFOWKTM (CC BY 3.0).

which was intended to draw water from hand-dug wells on farms. The water was then stored in redwood holding tanks.[8] These systems provided a self-contained water system, appropriate for new settlements that were isolated from rivers and other surface water sources. Later, the windpumps were constructed of steel, with more blades to provide increased torque for pumping. Finally, many were adapted for producing electricity. At their peak in 1930, there were an estimated 600,000 electric-producing windpumps in use in the US with a capacity equivalent to 150 megawatts.[9] Many of these settlements lacked electricity; by the mid 1930s, President Franklin Roosevelt's Works Progress Administration (WPA) and the Rural Electric Administration (REA) prioritized rural electrification. Ultimately these programs contributed to the windpump's demise. Still, many windpumps continue to pump water today.

Windpumps did suffer from mechanical efficiency problems and the inconsistencies of the weather, which could be overcome with electricity. At the same time, they don't consume fossil fuels. More recently, researchers around the world have begun to explore improvements to windpumps for reasons of rising energy costs and sustainability goals. Beginning in the 1980s, the US Department of Agriculture (USDA) and Department of Energy (DOE) sponsored research efforts to improve the efficiency of these devices.[10] While interest has remained strong in finding uses for this passive technology, for pumping water or other purposes, several manufacturers now offer smaller personal wind turbines which can generate a modest amount of electricity especially for those living in remote areas or far from the electric grid.

Wind Turbines

Modern wind turbines represent a continuation of thousands of years of windmill and windpump development, although they differ by generating electricity on a utility scale. Numerous early attempts failed to convert principles of windmill and windpump design to electricity generation. In 1887, English academic John Blythe constructed what some consider to be the first wind turbine for his country home in Scotland.[11] Around the same time, American Charles Brush devised a design made of metal (resembling a big windpump) which was constructed in Cleveland, Ohio also in 1887.[12] Measuring about 60 feet tall, it was never economical to use but foreshadowed, like Blythe's turbine, what was to follow a hundred years later. Between the 1930s and 1950s, some of the first documented, operable wind turbines were connected to an electricity grid in the USSR, UK and US.[13] By the 1980s, amidst concerns about oil supplies and the safety of nuclear power, several countries initiated plans to develop the technology again. Today, Denmark and China lead in the manufacture and use of wind power technology. Denmark, a country of 5.8 million, has set ambitious plans to generate all electricity and heat using renewable resources by 2035.[14] Presently, the nation generates roughly 45% of their electricity from wind. Other countries have set similar goals in coordination with the United Nations Sustainable Development goals and the Paris Climate Treaty. Estimates vary, but average wind turbine electricity generation has increased roughly 50 times since 1990 as the blade design and mechanics have improved. Looking ahead, cost efficiencies continue to encourage the adoption of wind power, over fossil fuels and alternatives, to power lighting, transportation and other modern necessities.[15] Wind power has also been met with criticism for a variety of reasons, including visual intrusiveness and harming wildlife. In 2021, severe cold

weather in Texas shut down wind turbines (and natural gas pipes), leaving millions of customers without heat and electricity for days. Critics of wind power seized the opportunity
to undermine the technology.[16] Experts countered that Texas' turbines failed because they
were insufficiently winterized. Despite this passing controversy, wind remains an underused, renewable global resource. With natural gas shortages triggered by sanctions after
the Russian invasion of Ukraine in 2022, the development of this technology has become
ever more urgent. As wind power continues to grow as a share of global electricity production, some new turbines have even been proposed that could harness fierce storm
winds in hurricane and typhoon zones.[17] However near-term these proposals are, wind
power undoubtedly merits committed investment alongside solar, wave and other energy
harvesting technologies.

Figure 9.5
**An 1891 image of Charles Brush's early wind turbine. Note the human at bottom right for scale. Source:
Photographer unknown**[18]

Manufacturing with Water

Scholars generally agree that water-powered mills and devices developed independently
in ancient Greece, the Roman Empire and China around two thousand years ago, perhaps
earlier.[19] Applications for these technologies have been vast, ranging from grain, lumber
and industrial production. Just as wind can be converted mechanically for the creation of
electricity, water power technology followed a similar trajectory. Water-powered turbines
were and continue to be used today, mostly at the utility scale. For a short period at the

turn of the 20th century, water was also used to power home appliances before widespread home electrification.[20] This type of hydropower called "direct hydropower" is of course distinct from the large-scale hydropower infrastructure projects from the Hoover Dam in the US (1936) and the Three Gorges Dam in China (2003). In the US, many of these aging larger hydro-electric dams are now being dismantled, often to help declining fish populations.[21]

Water mills were once a principal means of harnessing mechanical power, but have become less common. However, recent research projects have illustrated that such direct hydropower projects could (and still do) provide reliable, local sources of mechanical energy and electricity. One such project, led by Dr. Brian Raichle at the Appalachian State University, demonstrated this possibility. Raichle created a direct "Micro-hydro"-powered coffee de-pulper in Nicaragua, to support the processing of this regionally vital cash crop. In this instance, the system was constructed using the mechanical motion from a source of moving water.[22] Roughly a hundred years ago, direct hydropower was also used in European and American homes to produce auxiliary electricity.[23]

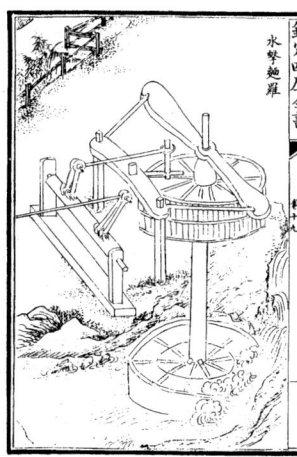 

Figure 9.6
Above left: A water piston invented by Chinese mechanical engineer Wang Zhen in the early 14th century. Source: Gisling modification of drawing by Wang Chen, 1290–1333.

Figure 9.7
Above right: A series of hydraulic turbines used to electrify the Roanoke Electric Railway, early 20th century. Source: Special Collections, Virginia Tech University Libraries.

The Society for Establishing Useful Manufactures (SUM)

In the early years after the United States became an independent nation, Alexander Hamilton, the first secretary of the treasury, initiated plans for America's first industrial center in 1791. The plant was noteworthy in that it was entirely powered by water. With the financial support of investors, Hamilton sought to achieve industrial independence from manufacturers in the UK. He found a suitable location to establish the direct hydro-powered industrial complex in Patterson, New Jersey, where there was access to natural resources, a sizable population and most importantly, the Passaic River's Great Falls, one of the largest waterfalls in the eastern United States. $500,000 was secured to develop the complex, an enormous sum for that time.

A charter granted by the New Jersey State legislature gave the Society for Establishing Useful Manufactures (SUM) perpetual exemption from county and township taxes and gave them the right to hold property, improve rivers, build canals and raise additional financial resources. Hamilton then commissioned Frenchman Pierre Charles L'Enfant, the celebrated architect and urban planner for Washington D.C., to layout the town of Paterson and build raceways to supply water to the site of one of the first waterpowered industrial mills in the United States. L'Enfant had a penchant for grand schemes and proposals. He visualized a series of radiating roads for the town (recalling the plan for Washington D.C.). The water-power system would have a grand aqueduct that would channel water from the Passaic River, across a ravine and around a hill to the mills. This aqueduct would be large enough for barge traffic and have a towpath on one side and a carriageway on the other. He also planned a reservoir for water storage, in essence an early battery technology, to ensure an adequate supply for the mills.

The SUM replaced L'Enfant early on because his plan was considered too elaborate. In his place, the SUM hired the more pragmatic Peter Colt to implement a more straightfor-ward scheme.[24] Colt simply dammed up the ravine at one end, creating a reservoir, and sent the water into a single raceway down the hillside to the mills. In these mills, textiles, tools, firearms and other manufactured goods were produced to meet the needs of the newly independent nation especially after trade had been largely cut off from the UK after the American Revolution. Remarkably, this hydro-powered mill remained operational through the first half of the 20th century before mills running on coal, petroleum and later electricity prevailed over the aging hydropower complex. Elsewhere, hydro-powered mills flourished until the early 20th century in the US, Europe and elsewhere.

Manufacturing Goods with Solar Power: Solar Mill

Solar photovoltaics (PVs) were first developed in 1839 by Alexandre Becquerel, a French physicist who discovered that sunlight could be converted into electricity using metal electrodes and electrolytes. Over the following 150 years, scientists experimented with different materials and chemistries allowing the PV cell to slowly increase in efficiency.[25] Since the mid 1970s, PVs have advanced in conversion efficiency, increasing roughly by a factor of three, from 16% efficiency to as much as 47%.[26] Their costs have simultaneously declined. Systems and products employing photovoltaics have steadily risen as these efficiencies and economies of scale have enabled them to compete with other forms of energy. Circular Economy specialists often speak of "Embodied Energy" in making indus-trial products, or the amount of energy used to manufacture products in their entirety from mining materials, shipping to the energy used in manufacturing from toothbrushes to automobiles and buildings.[27] To what extent the energy source is renewable is also con-sidered. Today, ecologically minded companies often promote their carbon neutrality, often because they buy grid electricity generated from solar or wind power instead of fossil fuels to make their products.[28]

Embodied energy to manufacture goods has long been provided by wind, water or other naturally harvested sources. Now that PVs have become more economically viable and efficient, efforts to build products and factories using machinery powered entirely by PVs is underway, including Tesla's battery factories. One smaller example of these efforts is SolarMill, a Richmond, Virginia-based contract manufacturer. SolarMill offers contemporary fabrication services: computer numeric controlled (CNC) machining, routing, laser cutting and other fabrication offerings to local and national clients. Like earlier water- and wind-powered machinery, SolarMill's energy budget is tied to the electricity levels of a battery bank, charged by an array of PVs on the roof of their plant. Just as manufacturing operations used to be tied to wind availability and water flow using wind and water mills, SolarMill makes production scheduling decisions based on weather forecasts and remaining battery levels. SolarMill has also developed software which can calculate an "energy budget" for a given design. Energy is stored in a battery bank so it can be used on demand.[29]

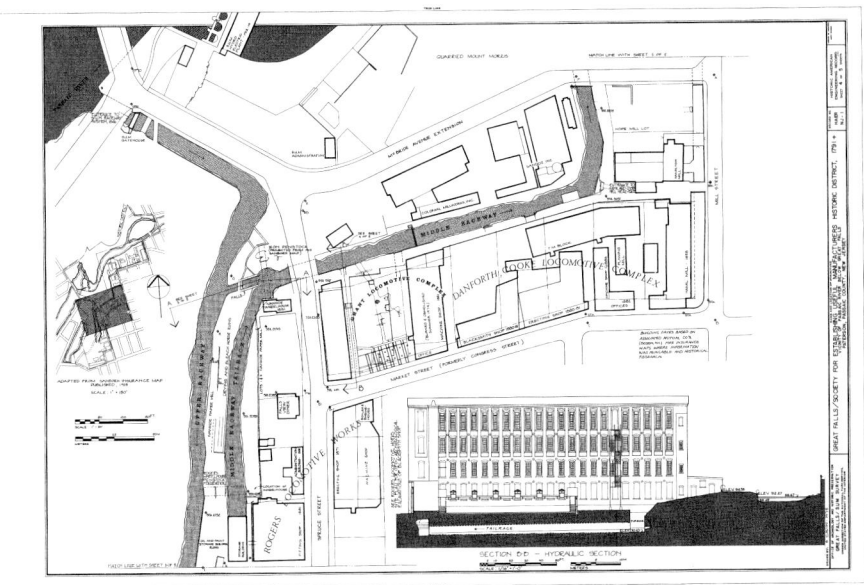

Figure 9.8
A 1975 Historic American Engineering Record drawing of the Society for Establishing Useful Manufactures (S.U.M.) complex in Patterson, NJ, USA showing water raceways and manufacturing plants. Founded by Alexander Hamilton to encourage domestic manufacturing before the widespread adoption of steam power. S.U.M. operated in waves from 1794 through its dissolution in 1945. Source: Library of Congress, HAER, B. Cavin, 1975.

Pedal-Powered Tools: Sewing Machines, Lathes and Home Appliances

Smaller industrial machines relying on human power were also used in the 19th century to help fabricate products on an industrial scale ranging from garments to furniture. While labor conditions in these early factories were often very poor, some of these manual tools could be reimagined for ethical production today or for personal use in the home.[30] By the turn of the 20th century, electricity became available to power industrial machines, either through a central motor driving individual machines with mechanical belts or later through dedicated electric motors for each machine. While human power persists today with bicycles, few examples of human-powered machines have endured, in industry and for home use. This could change depending on the scale and goals of the individual companies that would use them. Perhaps one of the most recognizable examples of manually powered fabrication tools is the original foot-powered sewing machine, introduced in the 19th century. Whether used in a garment factory or at home, the operator would use modest leg motion to drive the sewing machine's mechanics. Similar tools turned wood parts on lathes or performed other machining, weaving or fabrication tasks. Sometimes flywheels would be attached to develop enough inertia to undertake heavier duty operations. Most of these machines used heavier materials of the time, including wood and cast iron. Today, with the availability of harder and lighter materials, perhaps human effort could be partially restored to some of these production processes, to reduce embodied energy, just as we have seen with the Fiskars push lawn mower and concept carpet sweepers, balancing energy creation with some light physical activity. Indeed, many of these mechanical motions, through pedaling or other forms of physical activity, are used today in exercise gyms for physical fitness. What if these exercise regimens could be harnessed to fulfill some of societies power needs? Writer and researcher Kris De Decker and artist Melle Smets explored these questions, through their proposal *Human Power Plant*. Together, they envisioned converting an empty dormitory in the Netherlands into a community powered entirely by manual machines that would double as fitness equipment. This effort, they believe, could help TU Delft reach their carbon neutral goal by 2030.[31] It is likely easy to be skeptical that human power could drive present industrial activity, let alone all of society's energy needs. But maybe it could play a greater role or at least an *educational* role—by raising awareness, and thoughtfulness, about the *connection* between human effort and energy consumption.

Figure 9.9
A foot-powered sewing machine by Singer. Models like these were used in the home and for industrial production using a treadle mechanism. Additional tools using foot power were also made and used successfully before electrification: Lathes, punches, among many others. The earliest SUM factories used similar machines powered by direct hydropower from Patterson Falls raceways. Source: Burma (CC BY 2.0).

Pedal Power and Global Economic Accessibility

Despite their obscurity in wealthier economies, some of these manual human-powered technologies have found practical applications in countries lacking sufficient energy infrastructure. Sometimes called Appropriate Technology, Intermediate Technology[32] or "Design for the other 90%,"[33] numerous examples exist. For instance, in 1991, Dr. Martin Fisher and Nick Moon established ApproTEC (now Kickstart International), a non-profit organization centered on empowering farmers with the tools to be self-sufficient. In Kenya, they developed the *Moneymaker* treadle pump to draw ground water for irrigation. Resembling a Stairmaster exercise machine, they are constructed of local materials and assembled and maintained by local tradespeople. To date, the company continues to quietly improve the livelihoods of millions in Kenya, Tanzania, Zambia and elsewhere.[34,35]

More recently, in 2010, University of Sheffield engineering student John Leary created Bicibomba Movil, a bicycle-powered water pump for irrigation and water distribution in Guatemala. Later, the NGO Maya Pedal learned of the invention and assisted in its distribution, with the goal of improving the daily lives of locals without unreliable (and expensive) fossil fuel machines.[36] The product works by plugging a normal bicycle's rear wheel to a friction drive, connected to a mechanical pump. The back tire is thus connected to the pump mechanism, and powered by pedaling. The machine can pump 40 liters of well water per minute, and its design is prized by locals for its portability (after pumping, the frame can be pivoted upside down to sit on top of the rear wheel, so the bike can travel to the next well). The pumps remain available today through open-source and online DIY resources.

Notes

1 Historically, slaves have been forced to perform laborious tasks. This book acknowledges and condemns these injustices, focusing instead on human powered tasks performed by fairly paid laborers.

2 Hills, R.L. (1996) *Power from wind (a history of windmill technology)*. 1st edn. Cambridge: Cambridge University Press. ISBN 052156686X.

3 Hassan, A.Y. and Hill, D.R. (1986) *Islamic technology: An illustrated history*. Cambridge: Cambridge University Press. ISBN 0-521-42239-6.

4 Adam, L. (2006) *Wind, water, work: Ancient and medieval milling technology*. Leiden: Brill Academic Publishers. ISBN 90-04-14649-0.

5 US Energy Information Agency (2022) *History of wind power*. Available at: www.eia.gov/energyexplained/wind/history-of-wind-power.php (Accessed: August 11, 2022).

6 Kettering, A.M. (2008) 'Landscape with sails: The windmill in Netherlandish prints,' *Simiolus: Netherlands Quarterly for the History of Art*, 33(1/2) pp67–80. Available at: www.jstor.org/stable/20355351 (Accessed: August 11, 2022). "The Mill" painted between 1645–1648 by Rembrandt Van Rijn presents the windmill as an heroic national icon of productivity.

7 De Decker, K. (2009) 'Wind powered factories: History (and future) of industrial windmills,' *Low-Tech Magazine* [online]. Available at: www.lowtechmagazine.com/2009/10/history-of-industrial-windmills.html (Accessed: August 11, 2022).

8 Halladay, D. (1854) *Wind wheel* [online]. Available at: https://patents.google.com/patent/US11629A/en (Accessed: August, 11 2022).

9 Gipe, P. (1995) *Wind energy comes of age*. 1st edn. New York: John Wiley and Sons. ISBN 0-471-10924-X.

10 Hagen, L.J. and Sharif, M. (1981) 'Darrieus wind turbine and pump performance for low-lift irrigation pumping.' Prepared by the U.S. Department of Agriculture Agricultural Research Service. Prepared for

the U.S. Department of Energy Federal Wind Energy Program [online]. Available at: https://infosys.ars. usda.gov/WindErosion/publications/Andrew_pdt/79-345-D.pdf (Accessed August 11, 2022).

11 Price, T.J. (2004) 'Blyth, James (1839–1906),' *Oxford dictionary of national biography* (online ed.). Available at: www.oxforddnb.com/view/10.1093/ref:odnb/9780198614128.001.0001/odnb-97801986 14128-e-100957 (Accessed: August 11, 2022).

12 Righter, R.W. (1996) *Wind energy in America: A history.* Norman: University of Oklahoma Press, p44 ISBN: 0806128127.

13 Wyatt, A. (1986) *Electric power: Challenges and choices.* Toronto: Book Press. ISBN 978-0-920650-01-1.

14 Go100% (n.d.) '100% Renewable Denmark.' Available at: www.go100percent.org/cms/index.php? id=70&tx.ttnews%5Btt_news%5D=109 (Accessed: August 11, 2022).

15 Fares, R. (2017) 'Wind energy is one of the cheapest sources of electricity, and it's getting cheaper,' *Scientific American*, August 28 [online]. Available at: https://blogs.scientificamerican.com/plugged-in/wind-energy-is-one-of-the-cheapest-sources-of-electricity-and-its-getting-cheaper/ (Accessed: August 11, 2022).

16 Douglas, E. and Ramsey, R. (2021) 'No, frozen wind turbines aren't the main culprit for Texas' power outages,' *Texas Tribune*, February 16. Available at: www.texastribune.org/2021/02/16/texas-wind-turbines-frozen/

17 Murayama, R. and Sheldrick, A. (2021) 'This Japanese start-up has designed a wind turbine that can work in typhoons,' *World Economic Forum*, November 25 [online]. Available at: www.weforum.org/agenda/2021/11/japan-start-up-wind-turbine-typhoons-energy/ (Accessed: August 11, 2022).

18 Unknown author. Source of image: Robert W. Righter (1996) *Wind energy in America: A history.* University of Oklahoma Press, p44. Retrieved on December 27, 2008. ISBN: 0806128127.

19 Wikander, Ö. (2000) 'The water-mill,' in Wikander, Ö. (ed.) *Handbook of ancient water technology. Technology and change in history, vol. 2,* Leiden: Brill, pp. 371–400. ISBN 90-04-11123-9.

20 Hunter, L.C. and Bryant, L. (1991) 'A history of industrial power in the U.S., 1780–1930: Vol 3: The transmission of power first edition,' MIT Press, Cambridge. (US) First Printing ISBN: 9780262081986.

21 Flaccus, G. (2022) 'Federal report boosts plan to remove 4 dams on Klamath River' [online]. Available at: www.opb.org/article/2022/08/26/klamath-river-dam-removal-coho-salmon-migration-oregon-dams/ (Accessed: August 11, 2022).

22 Raichle, Brian W., Sinclair, Raymond S. and Ferrell, Jeremy C. (2012) 'Design and construction of a direct hydro powered coffee depulper,' *Energy for Sustainable Development*, 16(4), pp401–405. ISSN 0973-0826, https://doi.org/10.1016/j.esd.2012.08.006

23 *Low-Tech Magazine* (2013) 'Power from the tap: Water motors,' September 9. Available at: www.lowtechmagazine.com/2013/09/power-from-the-tap-water-motors.html#more (Accessed: February 3, 2022).

24 Fries, R.I., Cavin, B., Young, J. and Kramer, M.E. (1977) 'The Great Falls raceway and power system, Paterson, N.J.' Dedication by the American Society of Civil Engineers [online]. Available at: www.asme.org/wwwasmeorg/media/resourcefiles/aboutasme/who%20we%20are/engineering%20his tory/landmarks/28-great-falls-raceway-and-power-system-1792.pdf (Accessed: August 11, 2022).

25 Han, A. (2020) *Efficiency of solar PV, then, now and future* [online]. Available at: https://sites.lafay-ette.edu/egrs352-sp14-pv/technology/history-of-pv-technology/ (Accessed: August 11, 2022).

26 National Renewable Energy Laboratory (2022) *Best research-cell efficiency chart* [online]. Available at: www.nrel.gov/pv/cell-efficiency.html (Accessed: August 11, 2022).

27 MacArthur, E. (2020) *What is a circular economy?* [online]. Available at: www.ellenmacarthurfounda tion.org/circular-economy/concept (Accessed: August 11, 2022).

28 Visram, T. (2022) *Climate neutral is trying to build a net-zero labeling system that drives change-and dollars* [online]. Available at: www.fastcompany.com/90773451/climate-neutral-are-is-trying-to-build-a-net-zero-certification-system-that-drives-change-and-dollars (Accessed: August 11, 2022).

29 Hutson, M. (2022) 'Potential energy: The rush to develop renewable storage,' *The New Yorker*, April 25 [online]. Available at: www.newyorker.com/magazine/2022/04/25/the-renewable-energy-

revolution-will-need-renewable-storage (Accessed: August 11, 2022). New technologies, based on past inventions, are now exploring ways to store potential energy, like batteries, except with water, sand or other materials placed at a higher elevation. When electricity is needed, the materials are released and through gravity they turn turbines which generate electricity.

30 Chop with Chris (2013) *The foot powered (treadle) lathe* [video file], April 11. Available at: www.youtube.com/watch?v=eG9R0q9QJQc

31 De Decker, K. and Smets, M. (2017) 'Could we run modern society on human power alone?' *Low Tech Magazine* [online]. Available at: www.lowtechmagazine.com/2017/05/could-we-run-modern-society-on-human-power-alone.html and www.humanpowerplant.be/human_power_plant/human-powered-student-building-plans.html (Accessed: August 11, 2022).

32 Schumacher, E.F. (2010) *Small is beautiful*. New York: Harper Collins.

33 Smith, C. (2007) 'Design for the other 90%,' *Cooper Hewitt* [online]. Available at: www.cooperhewitt.org/publications/design-for-the-other-90/ (Accessed: August 11, 2022).

34 Russel, R. (2004) 'Pumping prosperity,' *Stanford Social Innovation Review*, 2(3), pp51–52. https://doi.org/10.48558/2CHX-JM49 (Accessed: August 11, 2022).

35 Galvin, M.D. and Iannotti, L. (2015) 'Social enterprise and development: The KickStart model,' *Voluntas: International Journal of Voluntary and Nonprofit Organizations*, 26(2), pp421–441. www.jstor.org/stable/43654689

36 Leary, J. and Marroquin, C. (1997) 'Bicibomba Movil (Mobile Bicycle Powered Water Pump). Construction Manual' [online]. Available at: www.mayapedal.org/Bicibomba_Movil_eng.pdf (Accessed: February 3, 2022).

Chapter 10

Transporting Goods and People

Today, transportation consumes a significant percentage of global energy: as much as 25% of total energy consumption is devoted to the transport of people and goods.[1] In the US, this percentage is even higher, at 28%.[2] This chapter concludes the topics covered in this book by examining alternatives for the transportation of people and goods, building on the partial fate of electric surface transit discussed earlier in Chapter 2. Electric "interurban" rail, freight transport, canals and additional water-powered modes of transport, including reaction ferries, water-powered funiculars and others, will be explored.

Electric Interurban Passenger Rail

In addition to the well-known electric streetcar systems found in cities worldwide, regional electric interurban trains once provided slightly longer distance transportation, between nearby city centers of all kinds, particularly in the US, before the arrival of the automobile. Similar kinds of train lines exist in Japan, Switzerland, the Netherlands and elsewhere.[3] Distinct from commuter train lines running from rural and suburban outskirts into urban city centers, interurbans were designed to connect cities, sometimes smaller ones and often the remote rural communities in between. Trainsets tended to be faster than urban streetcars but lighter than conventional railroads. Nevertheless, they would still provide passenger service to riders in either direction like suburban commuter trains. Some would also provide regional freight service. At the height of their use, as many as two hundred interurban systems were operated across North America and dozens more were planned, partially built but never completed. Overall, interurbans provided a relatively quiet, fast and efficient means of regional travel. Some of the last surviving interurbans were taken out of service as recently as the early 1960s, notably the Pacific Electric Railway's Los Angeles to Long Beach line (service was restored in 1990 as a part of the Los Angeles Metro). Many others ceased operations in the years following the Second World War as roads, the automobile and other travel options expanded. Today, a few lines have survived: Philadelphia's Route 100 to Norristown, and Chicago's South Shore Line to South Bend, Indiana. Many of these interurban trains travel at speeds higher than current highway speed limits. SEPTA's Route 100's former "Bullet" cars, for instance, built in the early 1930s, regularly exceeded 80mph. Made of lightweight aluminum and powered by four 100 horsepower electric motors, the Bullet cars were some of the first vehicles to be developed and tested in wind tunnels.[4] Part of what allowed these trains to reach these speeds was their infrastructure. Unlike many railroad systems even today, Route 100 has no traffic crossings on the entire 13.4-mile line.

DOI: 10.4324/9780367814304-11

Figure 10.1
A 1907 network map of the Pacific electric interurban system of Los Angeles. Difficult to believe now but LA's interurban system, traveling between regional urban cores, was the envy of the nation. Today, some of the system has been revived but with limited success, in part due to LA's urban decentralization, born in part after decades of car-based urban development. Source: mark6mauno (CC BY 2.0) Cropped by Brook Kennedy.

Each car could carry roughly 50 passengers. In 1990, after 60 years in service, the 11 car fleet of Bullet cars was retired, replaced by modern equipment, and the line remains well-used today. Two other interurban lines in central New York and Utah used the Bullet cars but ceased operations in the 1930s and 1950s respectively.

Unlike Route 100, the South Shore Line in Indiana fluctuated between a separate private right of way and one that until recently, ran down main streets in Michigan City, Indiana—a remnant of a pre-automobile time. Whether speaking of electric interurbans or commuter train lines that were discontinued, both offered a vital service to metropolitan and rural communities—fast, efficient, electric transportation options.[5]

Today, there has been renewed interest in building interurban train services using former and disused rail lines. For example, the New Jersey Transit "River Line" was constructed in 2004 between Trenton and Camden on underused freight tracks. Running on diesel power rather than electricity, it connects two smaller cities in the state and provides rail commuting options for those living in between. Unfinished proposals have been made to create longer high-speed intercity lines in California and Texas—to relieve traffic congestion, pollution and carbon emissions. Acela, the high-speed intercity rail system introduced in the Northeast US in 2000, continues to attract strong ridership much like high-speed rail systems around the world. Newer technologies, partially inspired by interurban railways, like Hyperloop have also been proposed with the same goal: to deliver high-speed, energy-efficient travel between population centers. In Silicon Valley, the nexus of new technology, the local diesel Caltrain commuter line from San Francisco to San Jose (and beyond) is currently being electrified, demonstrating that interurban rail has relevance in current transportation ecosystems especially where cities and major population centers are clustered. During the Covid-19 pandemic, regional air travel (and rail) plummeted as travelers feared crowded sealed cabins. Yet people still need to get around locally and desire alternatives to driving. Some have argued that fast electric regional rail and/or buses could compensate for shorter distance regional air travel.[6] At the same time, such a transition could meaningfully reduce energy consumption, emissions (and stress) associated with regional air travel.[7] In the summer of 2022, Germany offered subsidized monthly rail tickets for €9 to help ease the burden of inflation and rising energy costs. Over the course of three months, 52 million tickets were purchased, and the reduction in car use has already resulted in decreased air pollution and carbon dioxide emissions.[8]

Electric Freight Rail

Like many kinds of electric surface transit systems, electrically powered freight rail has also been mostly abandoned in the US. While using electricity to transport goods by rail persists in Europe and Asia, it has mostly vanished in the US with the exception of a few small freight lines. Electric freight rail has been reconsidered recently in battery-electric renditions by General Electric (GE) and Burlington Northern Santa Fe (BNSF), and has been a part of grand federal infrastructure-making programs to reduce emissions, especially in urban areas. Originally, it was also local government that led to the development of electric freight rail in the first place, over one hundred years ago. Then, laws were enacted to prohibit the use of steam power in tunnels, as the fumes and lack of ventilation were considered then, as they are now, hazardous to health. Maryland was one of the first states to pass such laws—as a result, the Baltimore and Ohio railroad developed electric locomotives to drag steam locomotives (and their freight) through tunnels beginning in 1895. New York City passed similar laws in 1903, which led to the construction of train tunnels under the Hudson River and Park Avenue in anticipation of the building of both the current Grand Central and former Pennsylvania Stations.[9] Today, the longest and one of the busiest freight tunnels in the world, the Gotthard Base Tunnel, passes under the Alps in Switzerland. At 35.5 miles long, passage would not be possible with steam and even diesel power, without significant ventilation interventions.

THE NEW ELECTRIC LOCOMOTIVE OF THE BALTIMORE AND OHIO RAILROAD COMPANY

Figure 10.2
An illustration of a Baltimore and Ohio electric-powered locomotive pulling a freight train with steam locomotive attached through a tunnel near Baltimore. Electric power was once the only option when steam was banned from underground use. Diesel locomotives are now permitted underground and have replaced electric freight rail in the US. Source: *Scientific American*, 10 August, 1895.

After the adoption of electric freight rail, its use grew in the early 20th century in the US, peaking in the 1930s as steam lines converted, then declined steadily through the present as diesel power prevailed. Over time, freight lines that were originally electrified later converted to diesel propulsion, some, notably the Virginia Railway, using electricity into the late 1950s. One of the reasons for the success of diesel is lower upfront costs: electrification systems, including overhead wire or third rail infrastructure, is costly for private companies to finance. Diesel locomotives, on the other hand, remain more expensive to maintain and are less energy efficient to operate.[10] In contrast to the US, most freight systems around the world are state sponsored, enabling the investment in the needed infrastructure. Notwithstanding, diesel freight rail has been prospering for decades alongside trucking. Compared with truck-based freight transport, train infrastructure offers advantages in energy efficiency. First, average sized freight trains consume less energy per weight hauled than an average semi-truck.[11] Additionally, steel rails create less friction than roads compared with that of a rubber tire on asphalt.[12] Rail right of ways are also more level, passing through short hills rather than climbing over them, as trucks do with the rest of vehicular traffic.

Even though studies have demonstrated the efficiencies of rail freight compared with individual diesel trucking, electric rail would provide further energy savings and opportunities to remove harmful emissions from urban areas.[13] Studies have proposed regional electrification of freight rail serving the ports of Los Angeles and Long Beach, the busiest in the US, to help reduce emissions and improve air quality. Options would include overhead catenary wire electrification and newer battery electric technology. In 2021, a legislative proposal was introduced to electrify much of the US's freight rail network—a project on the scale of Presidents Dwight Eisenhower's interstate highway system in the 1950s or Franklin D. Roosevelt Works Projects Administration (WPA) of the 1930s. The likelihood that such a project would be carried out is less a matter of desirability and more a question of cost and taxpayer acceptance. Some studies have also estimated that shipping costs and pollution could be reduced in the long term.[14]

Electric-Powered Delivery of Goods

As the popularity of online retailing and home delivery of goods has increased, so-called "last mile delivery" vehicles fulfilling these orders have proliferated. In contrast to today, and going back in time to the early 20th century, considerable volumes of freight were carried by electric-powered freight locomotives often on the same interurban tracks and infrastructure used to carry passengers. Perhaps inspired by these forms of freight transport, urban planners along with transportation design specialists PriestmanGoode envisioned "Metro Freight," a concept to adapt urban metro systems to deliver goods.[15] Taking advantage of lulls in peak commuter service, particularly overnight when many metro systems go out of service, PriestmanGoode proposed that goods could be loaded onto dedicated or existing metro rail cars at night when public transport is furloughed or running at reduced capacity. Goods could then be transported on these public trains and taken to various metro stations around the city—as close as possible to the final delivery location. Platforms at these metro stations could be adapted to facilitate the loading and unloading of goods, which could then be transferred to a collection point at the station. Consumers would then have the option to pick up goods directly from their local station and then continue on foot, bicycle, scooter or car.[16]

Gray Water-Powered Hill-Climbing Trains: Reflections by Dieter Rams

Electricity is, of course, not the only means of powering the transportation of people and goods. Passive flow of water and wind has also been used to power vehicles of all kinds. In Gary Huswitt's 2018 documentary *Rams*, the Industrial Design maestro Dieter Rams reflects about his boyhood when he used to ride a funicular train to climb a steep hill in a park in his hometown Wiesbaden, Germany.[17] Composed of two cars, one that climbs while the other descends, they are tethered with a cable. Without using electricity, the uphill car has a tank which is filled with water to make it heavier than the car and passengers below. Once the heavier uphill car pulls the lighter car up and reaches the bottom, it releases the water and the process begins again. Rams characterized the antique train as an early example of "ecological design." A funicular is essentially a parallelogram-shaped train that scales a steep incline, sometimes with two cars that use the principle of balance to bring passengers and freight both up steep hills and down. Many of these trains are used around the world in urban areas with steep terrain. From Valparaiso, Chile, Hong Kong, Pittsburgh, Pennsylvania and Switzerland, this technology did and continues to serve a vital service in these locations. Valparaiso's funiculars are also a UNESCO World Heritage Site. Today, 16 out of more than 25 funiculars still exist, only seven of which continue to operate in 2019. Switzerland, a country of steep mountain valleys and a pioneer in developing this technology, has the most working lines intact. One example in Fribourg further decreases energy use by harnessing waste water from homes and businesses to fill the balancing tank. Built in 1899, it is used by locals and tourists. Unlike other water-powered funiculars, Fribourg's funicular does not need to pump water back up the hill. That a train could carry 20 passengers up a steep hill without using fuel is not only a remarkable achievement in its time; it would still be today. Perhaps similar modern funiculars could be built in the steepest hills of Pittsburg, Pennsylvania and San Francisco, California today, like the three surviving cable car lines that draw tourists to the city already.

Figure 10.3
A waste water-powered funicular train in Fribourg, Switzerland. Source: Benediktv (CC BY 2.0).

With the heightened regional interest in "green energy" and "carbon neutrality" in California, such a proposal would provide interesting contrast to the onslaught of electric cars and driverless innovations. Perhaps new water-powered funicular trains could be optimized to accommodate bicyclists as well, to help riders cope with the steepest of hills in their morning and evening commutes. Modern versions of these trains could have entirely autonomous valve operation, braking and fare management while harnessing the natural flow of water to pull the weight of passengers uphill.

Reaction Ferries: Harnessing Water Current for River Crossings

In addition to moving passengers up and down steep inclines, several countries in Europe, New Zealand, Canada and the US have surviving remnants of another technology: reaction ferries. Using the natural force of water flowing downstream, reaction ferries traverse rivers held on course by a cable either overhead or on the surface of the water. In the city of Basel, Switzerland there are four surviving reaction ferries that cross the Rhine river. In this instance, the Basel ferries employ thin overhead cables to guide the ferryboats and their passengers across the river.

Figure 10.4

A reaction ferry crossing the river Rhine in Basel, Switzerland without using a paddle, motor or any energy. By harnessing the flow of the moving water, the ferry, held in place by a cable, crosses the river with passengers several times a day. Several reaction ferries operate in Germany, Poland, Italy and the UK. Some were operated in the Ozark mountains in the US, ceasing operations by the mid 20th century. Source: Christina (CC BY 2.0).

Waterways and Canals

Some of the earliest evidence of using rivers to transport cargo dates back more than 5,000 years to the Tigris and Euphrates in what is now present-day Iraq.[18] In more recent history, long before railroads and trucks, barge canals were used in Europe and later in the northeastern US to transport goods and sometimes people, especially connecting major rivers where additional transportation options were available. The main advantage of waterways and canals over road or surface transport comes down to friction—it requires less force to move boats over water than carts over dirt roads. Some of the earliest canals were constructed in France, England and Denmark, but today, canals are found on most continents.

Early on they transported grain and agricultural products; later they moved coal and other fuels. Today many canals have been preserved as nostalgic engineering monuments for tourism. Even so, waterways remain an overlooked mode of efficient transportation.[19]

The Canal du Midi was one of the first known examples of a functioning canal in the modern sense, connecting the Atlantic Ocean via the Garonne from Toulouse to Étang de Thau feeding to the Mediterranean Sea. Roman emperors Augustus and Nero dreamed of connecting these large bodies of water for economic and other reasons. French king François I consulted Leonardo Da Vinci in 1516 to envision how such a large work of civil engineering could be implemented. By 1666 King Louis XIV committed financial resources to a proposal devised by engineer Pierre-Paul Riquet.[20,21] Less than two decades later, the Canal Du Midi opened to shipping traffic and remains open today, listed as a UNESCO World Heritage Site. The implementation of such a plan was an ambitious undertaking for the time as it would be now: 63 mechanical locks were constructed and powered passively with water diverted from the rivers—in its entirety, the canal was powered by water flowing down stream and horses or human power without fossil fuels of any kind until pleasure boating took over in the early 20th century.

Around the world and especially in the US, most canals have, of course, now been abandoned. One example, the Morris Canal in New Jersey, begun in 1826, transported coal, iron, grain, wood, bricks, hay, hides, sugar, lumber, lime, ice and numerous food products from eastern Pennsylvania to New York city. A series of locks and planes, all using direct hydro-powered turbines, delivered barges from sea level in New York harbor to a high point in central New Jersey, and then back down to the Delaware river. Between the locks and planes, mules pulled barges along towpaths.

Figure 10.5
Above left: A lock on the Morris Canal near Waterloo, New Jersey, US, showing the towpath on the left. Source: Library of Congress, Detroit Photographic Company.

Figure 10.6
Above right: A c. 1893 image of an electric-powered trolley boat on the Teltow Canal, Germany. With improving new battery technology, could our waterways be revamped and expanded where appropriate to use electric propulsion with batteries or electric lines? Source: Lämpel.

By the early 20th century, competition from railroads led to declining traffic on the Morris and others in the region resulting in their eventual closure in the early 20th century. Other canals fared better: the Erie Canal (now the New York State barge canal) connecting Lake Erie to the Hudson River remains open as a historic waterway and is mostly used by pleasure craft. Likewise, numerous canals in Europe, Canada and other continents remain open too, some still carrying freight traffic. Many of these canals have been carefully maintained and, in some cases, modernized with motorized propulsion systems. Similar to electric train technology, some used overhead catenary wires to drive electric tugs both on the water and alongside on the horse and mule towpaths.[22,23]

Today, waterway-based shipping remains a comparatively energy-efficient option for moving cargo, especially grain and bulk goods. A 2017 report by the National Waterways Foundation found that a typical US waterway barge could move a ton of cargo 647 miles on a gallon of fuel (547 miles with diesel).[24] This compares to 477 miles by rail and 145 miles by truck. Some believe that waterways could provide far more capacity if aging infrastructure were updated and repaired.[25] Yet despite the fuel savings of some waterway shipping, current watercraft rely on diesel propulsion with minimal emission control. Trans-oceanic container ships continue to burn heavy fuel oil, responsible for significant carbon and pollutant emissions.[26] As with other forms of transportation, electrifying these vessels will play an important role in reducing these environmental impacts. In the near term, it is more likely that ships traveling shorter and inland itineraries would be able to be fully electrified. For longer distances across oceans, additional propulsion opportunities are being explored today, including the use of sails to supplement engine power.

Sailing Into the Future? Possible Revival of Wind-Propulsion Technology

Sailing craft could provide a meaningful source of inspiration for reducing energy consumption along trans-oceanic navigational channels, especially along prevailing wind routes.[27] Sailing craft have navigated oceans for thousands of years, only to be abandoned for commercial use in the 19th century. Sailing craft are indeed more complicated, and their course is constrained by the direction of the wind. However, before steam power eclipsed the sail, sailing craft were the only means of transporting goods and people across oceans. Of course, sailing technology was also strategically significant for military dominance, colonization, fishing and trade, including the development of the slave trade. Sails were developed thousands of years ago, manifest in several different types of nautical designs. Early examples appear from the Middle East and Austronesia; later, Chinese and European variants followed. Today, sailing craft are primarily used recreationally, however some inventors have attempted to retrofit container ships with supplemental sailing rigging to help reduce fuel consumption. Some systems resemble multi masted, square rigged configurations of the 18th and 19th centuries, similar to the *Cutty Sark*, an iconic end-of-a-generation cargo sailing vessel built just as naval architecture began to adopt steam or hybrid propulsion.

Departing from the conventional layout of historical sailing craft, fledgling startups have produced sailing rigs that autonomously deploy and retract, to supplement engine power.[28] Resembling small spinnaker sails, these systems provide some reduction in fuel consumption without the obtrusiveness of a traditional sailing rigging which could interfere with loading and unloading cargo. Skeptics point out that overladen cargo ships paired with cheap fuel prices will support a status quo, however government regulators synchronized with emission control objectives (such as the Paris Climate Treaty) could incentivize shipping companies to explore alternatives including sails. Other types of cargo ships, including tankers and automobile carriers, have open decks and their contents can be unloaded from the side. Such configurations could offer possibilities for deck-top sail technology retrofits.

Figure 10.7
Above left: *Cutty Sark*, one of the last ocean-going sailing cargo vessels culminating from thousands of years of global sailing technology development. Source: Allen C. Green, State Library of Victoria.

Figure 10.8
Above right: The *Oceanbird* sail-assisted cargo ship concept. Several companies today are exploring ways to augment or replace fossil-fuel propulsion systems with traditional, emission-free sailing technology. Source: Oceanbird.

The Second Return of Electric Vehicles

In light of recent excitement about electric vehicles (EVs) beginning in the late 2010s, it is easily forgotten that a few earlier generations of electric cars were developed and subsequently abandoned. Before Tesla, Polestar and hybrid electrics from Toyota and GM, electric cars were popular around the turn of the 20th century in continental Europe, the US, the UK and elsewhere. Many of these early EVs, namely Detroit Electric and Baker Electric, were favored by the well heeled. At the time, cars traveled shorter distances and required frequent maintenance. While more expensive than their mechanical counterparts, early electric cars were preferred for their simplicity, relative quiet and lower maintenance as they are today.[29,30] About 100 years later, General Motors leased the EV1 which despite its popularity was canceled after a few years.[31] Now nearly 20 years later, EV sales are booming again after the successful introduction of the Tesla Model S, although battery supply chains have been challenged by limited supply of the precious metals needed to make them.[32]

Electric "Trolley" Trucks

Battery electric trucks, once produced by Edison Electric, were introduced at the same time as the first electric cars. Ideal for short distance hauls and deliveries, the technology languished in the 20th century except for mail, milk and other local forms of moving and delivering goods. Today, battery electric trucks have yet to compete with diesel power, but newer manufacturers Rivian and Arrival are building short distance "last mile" delivery fleets at the time of writing this book. Concurrent with Edison's efforts, many companies and townships developed electric trucks powered by overhead catenary wires similarly to electric surface rail systems. Today, some of these systems have endured.[33] In the Port of Los Angeles, in partnership with Siemens, Trolley Trucks have been piloted to reduce regional air quality problems.[34] While on the freeway they can charge their batteries and detach from the wires on demand. Abundant precedents for electric Trolley Trucks can be found in continental Europe, Russia and Ukraine. Overhead electric wire systems for common freight thoroughfares could prove useful as an enhancement in the emerging battery-powered EV landscape. Vehicles, especially trucks, could use the overhead wire power to recharge while driving on major roads leaving ports and distribution centers only to detach under battery power when completing a custom itinerary.

Figure 10.9
A freight trolley truck running on electricity from a catenary wire system used by the streetcar system.
Source: Kneiphof (talk | contribs) - CC-BY-2.0.

Notes

1 US Energy Information Administration (2016) *Transportation and energy consumption* [online]. Available at: www.eia.gov/energyexplained/use-of-energy/transportation.php (Accessed: August 4, 2022).
2 US Energy Information Administration (2021) *Energy use for transportation* [online]. Available at: www.eia.gov/energyexplained/use-of-energy/transportation.php (Accessed: August 4, 2022).
3 Steuart, W.M. (1905) 'Street and electric railways, 1902,' *Special Reports*, Bureau of the Census. Available at: https://archive.org/details/streetandelectr00censgoog/page/n5/mode/2up (Accessed: August 12, 2022).

4 Middleton, W.D. (1961) *The interurban era.* London: Forgotten Books.

5 Lipow, G. (2006) 'Rail freight is more efficient than truck freight,' *Grist* [online]. Available at: https://grist.org/article/freight-trains-19th-century-technology-due-for-a-21st-century-revival/ (Accessed: August 12, 2022).

6 Myerson, M.S. (2020) 'Replace short flights with buses,' *Politico* [online]. Available at: www.politico.com/interactives/2020/magazine-friday-cover-redesigning-the-world-coronavirus/ (Accessed: August 12, 2022).

7 Chiara, B.D., De Franco, D., Coviello, N. and Pastrone, D. (2017) 'Comparative specific energy consumption between air transport and high-speed rail transport: A practical assessment,' *Transportation Research Part D: Transport and Environment*, 52(A), pp227–243. https://doi.org/10.1016/j.trd.2017.02.006

8 Feingold, S. (2022) 'Germany's €9 transit ticket cuts 1.8 million tonnes of CO2,' *World Economic Forum*, August 31 [online]. Available at: www.weforum.org/agenda/2022/08/germanys-9-euro-transport-ticket-cut-1-8-million-tons-of-co2/?utm_source=linkedin&utm_medium=social_video&utm_term=1_1&utm_content=27172_Germany_trains_9_euro&utm_campaign=social_video_2022 (Accessed: September 12, 2022).

9 Wadsworth, G.R. (1905) 'Terminal improvements of the New York Central & Hudson River in New York,' *Railroad Gazette*, 39(366).

10 Nunno, R. (2018) 'Electrification of U.S. railways: Pie in the sky, or realistic goal?' *Environmental and Energy Study Institute* [online]. Available at: www.eesi.org/articles/view/electrification-of-u.s.-railways-pie-in-the-sky-or-realistic-goal (Accessed: August 11, 2022).

11 US Department of Transportation (1991) *Rail vs. truck fuel efficiency: The relative fuel efficiency of truck competitive rail freight and truck operations compared in a range of corridors* [online]. Available at: https://railroads.dot.gov/sites/fra.dot.gov/files/fra_net/16332/1991_RAIL%20VS%20TRUCK%20FUEL%20EFFICIENCY%20-%20THE%20RELATIVE%20F%20(2%29.PDF (Accessed: September 9, 2022).

12 Nunno (2018).

13 Nunno (2018).

14 Nunno (2018).

15 Hörl, B., Dörr, H., Wanjek, M. and Romstorfer, A. (2016) 'METRO.FREIGHT.2020 – Strategies for strengthening rail infrastructure for freight transport in urban regions,' *Transportation Research Procedia*, 14, pp2776–2784, Available at: https://www.sciencedirect.com/science/article/pii/S2352146516304859 (Accessed: August 11, 2022).

16 Maxwell, P. (2020) 'Why urban mobility's future is neither public nor private,' *Frame* [online]. Available at: www.frameweb.com/article/paul-priestman-interview-dromos (Accessed: August 11, 2022).

17 *Rams* (2018) Directed by Gary Hustwit [Film]. New York. Gary Hustwit.

18 Rodda, J.C. and Ubertini, L. (2004) *The basis of civilization – water science?* Wallingford, UK: IAHS Press.

19 Clowdis, C.W. and Horowitz, N. (2009) 'The river barge still plays a role in U.S. transportation,' *S&P Global* [online]. Available at: https://ihsmarkit.com/country-industry-forecasting.html?ID=106593483 (Accessed: August 11, 2022).

20 Riquet Bonrepos, P. (1805) *Histoire du Canal de Languedoc.* Paris: L'Imprimerie de Crapelet.

21 Gast. H. (2014) *Le Canal du Midi et Les Voies Navigables.* Rennes. Ouest-France.

22 De Decker, K. (2009) 'Trolley canal boats,' *Low-Tech Magazine* [online]. Available at: www.lowtechmagazine.com/2009/12/trolley-canal-boats.html (Accessed: August 11, 2022).

23 Editors (1894) 'Electricity on the canals; another scheme for utilizing the trolley system,' *The New York Times*, January 21. Available at: www.nytimes.com/1894/01/21/archives/electricity-on-the-canals-another-scheme-for-utilizing-the-trolley.html (Accessed: August 11, 2022).

24 Kruse, C.J., Warner, J.E. and Olson, L.E. (2017) 'A modal comparison of domestic freight transportation effects on the general public: 2001–2014,' *Texas A&M Transportation Institute* [online].

Available at: www.nationalwaterwaysfoundation.org/documents/Final%20TTI%20Report%20 2001-2014%20Approved.pdf (Accessed: August 11, 2022).

25 Waterways Council Inc. (2021) 'Inland waterways' [online]. Available at: https://infrastructurereport card.org/wp-content/uploads/2020/12/Inland-Waterways-2021.pdf (Accessed: August 11, 2022).

26 Walker, T.R., Adebambo, O., Del Aguila Feijoo, M.C., Elhaimer, E., Hossain, T., Edwards, S.J., Morrison, C.E., Romo, J., Sharma, N., Taylor, S. and Zomorodi, S. (2019) 'Environmental effects of marine transportation,' *World Seas: An Environmental Evaluation*, pp505–530. doi:10.1016/B978-0-12-805052-1.00030-9. ISBN 978-0-12-805052-1. (Accessed: August 11, 2022).

27 Peters, A. (2019) 'Cargo ships are big polluters. Can they go back to using sails?' *Fast Company*, July 17 [online]. Available at: www.fastcompany.com/90376983/cargo-ships-are-big-polluters-can-they-go-back-to-using-sails (Accessed: August 11, 2022).

28 Startup companies like Oceanbird and Skysails Marine have introduced self-deploying sails to help reduce fuel consumption. Several other startups have proposed various configurations to reintro-duce some sail power to the shipping industry.

29 Clymer, F. (1950) *Treasury of early American automobiles*. New York: McGraw Hill.

30 Barber, M. (2017) 'Before Tesla: Why everyone wanted an electric car in 1905,' *Curbed*, September 22, 2017. Available at: https://archive.curbed.com/2017/9/22/16346892/electric-car-history-fritchle (Accessed: July 19, 2022). Many early EVs also used Nickel-Iron batteries, which could last decades, and were easier to resuse and dispose. Their poorer performance in cold weather, amongst other concerns, have prevented their use in EVs today.

31 Taylor, M. (2005) 'Owners charged up over electric cars, but manufacturers have pulled the plug,' *SFGate* [online]. Available at: www.sfgate.com/green/article/Owners-charged-up-over-electric-cars-but-2677780.php

32 Rajaeifar, M.A., Ghadimi, P., Raugei, M., Wu, Y., and Heidrich, O. (2022) 'Challenges and recent developments in supply and value chains of electric vehicle batteries: A sustainability perspec-tive,' *Resources, Conservation and Recycling*, 180, p106144. https://doi.org/10.1016/j.resconrec.2021.106144

33 In Russia and Ukraine, electric delivery trolley trucks use electric streetcar catenary wires for the delivery of goods. Trolley trucks are also used extensively in mines in the US, Europe and Africa.

34 Roether, J. (2017) 'First electric highway in U.S. unveiled near ports of L.A. and Long Beach,' *Energized* [online]. Available at: https://energized.edison.com/stories/first-electric-highway-in-u-s-unveiled-near-ports-of-l-a-and-long-beach (Accessed: August 12, 2022).

Conclusion

Alternative Principles for 21st-Century Design

In the preceding chapters, this book examined a variety of alternative technologies: past, traditional and forgotten examples that could be reimagined for today, ranging in scale from urban design to hand-held tools. Undoubtedly, there are many more examples yet to be covered, all of which deserve closer evaluation. Hopefully more researchers and designers will build upon the work already begun here. Clearly, many examples of alternative technology, however old-fashioned some might appear, offer practical opportunities to design more responsibly, especially in accordance with the United Nation's Sustainable Development Goals: conserving energy, protecting natural resources, reducing carbon emissions and curbing pollution. To recall the goals presented in the introduction, the spirit of this book is to look forward, not to turn back the clock; to learn from alternative knowledge to design for the present—and the future. Looking back at the chapters that followed, notable patterns have emerged which should now be articulated and documented. These patterns are best summarized in four overarching themes or *principles*, in no order of priority and with some overlap. As designers envision the built environment in the 21st century, they should at least be considered.

Principle 1. Conserve Resources with Passive, Harvested and Manual Energy

As the human population grows, so too will the demand for energy, water and food.[1] At the same time, natural resources, including petroleum, timber and others, are expected to decline.[2] Where will all the necessary resources come from to power and furnish human lifestyles? Naturally harvested energy sources from solar and wind power have offered some optimism, but will unlikely satisfy projected global energy needs in the near future. As supply for fossil fuels peaks and demand rises, costs too will inevitably climb. Despite growing adoption of renewable energy, conservation will also be needed to address resource shortfalls. According to data gathered by the World Bank, Canada and the US, along with smaller petroleum rich countries like Bahrain, Qatar and Trinidad and Tobago, all lead in overall per-capita energy consumption. By comparison, the UK uses less than half of the energy of the US.[3]

Conserving energy can also be achieved by thinking *passively* or *manually* before using electricity or fossil fuels. This, of course, requires foresight, planning and design. In the past chapters, we have seen examples of energy harvesting funiculars, water pumps and ferries and broad passive and low-energy cooling techniques which can complement the use of air conditioning. Inside buildings, we have seen ways to introduce air flow and light without

DOI: 10.4324/9780367814304-12

flipping a switch. At the same time, this book has recognized that not all passive cooling solutions will likely be sufficient in the hottest climates when used alone, but they can be implemented in *hybrid* embodiments to reduce the overall energy consumption in buildings. Today, despite much excellent work to design NetZero buildings through Passive House and other energy efficiency standards, new commercial buildings, often in the US, leave individuals little agency to control the temperature (or energy use) of their personal work environments. Giving individuals the control to use passive alternatives, like an operable window on a cooler morning, or an external shade or even local climate control, can play a valuable role as well.

Designs based on manual power also provide benefits beyond energy and resource conservation—when powered systems fail, devices with manual options can still be used. Powered design that allows manual override, as discussed in Chapter 7, can provide continued use when batteries are depleted and sometimes fail. Opportunities abound to implement manual and manual override-based products for the home, work and transportation. E-bikes, as mentioned, can at least be pedaled when the batteries discharge, despite the extra weight of the electric drive system. They also empower the rider to not rely solely on electric energy to get where they are going for health benefits.

Principle 2. Balance Light Physical Activity with Power-Assist

Building on the theme of manual override, Principle 2 focuses on additional benefits, notably optional light physical activity. In the present era, for reasons of convenience and comfort, design in developed industrial economies have increasingly emphasized power-assisted operation, in as many areas of human activity where it can be applied. Whether it be hand-held blenders, wine bottle openers, automotive power windows or automatically closing tailgates, electrification and power-assist has taken over design to make life increasingly effortless. This has benefited those who are differently abled especially. However, inclusively designed automatic doors, specified by Americans with Disabilities Act (ADA) standards for the disabled, allow able-bodied individuals to enter buildings without using their arms or looking up from their smartphones. These effort-free values have also flourished through drive-through restaurants, banks, dry cleaners and curb service, ensuring customers never have to get out of their cars. Smartphones provide an additional layer: app-based services facilitate home meal delivery to avoid going out altogether.

Perhaps unsurprisingly, the very societies with plentiful access to basic resources like energy, water and food—the cornerstone of these minimal effort lifestyles—are often not the healthiest. The US, for instance, ranked number 46 in average life expectancy in 2020, while simultaneously leading in per-capita consumption of energy and natural resources, including pharmaceuticals.[4] The US also leads in average miles driven per day by car. No surprise again, Americans have some of the higher obesity rates in the world.[5] The World Health Organization recommends a combination of physical activity and diet to promote health, longer life expectancy and to reduce risk of heart disease, diabetes and other chronic disorders.[6] Yet the workforce of many industrialized nations is less physically active as jobs have transitioned from physical labor to more sedentary office work. This would clearly explain why the exercise industry has steadily grown each year and why designs like the standing desk and bicycling have become popular—to *reintegrate* physical activity into daily life.

While this book has already shown how manual products *en masse* can contribute to energy conservation in the home, it is not suggesting that manual products would replace

Figure 11.1
Top left: A Newton hand lever-powered espresso coffee maker. Source: Hayden, Newton Espresso.

Figure 11.2
Top right: Designer Antoine Pateau's hand-powered blender concept with swappable heads, "The Muscle the Gear and the Carrot." Source: Antoine Pateau.

Figure 11.3
Bottom: Manuel Immler's Pino hand-powered blender. Source: Manuel Immler.

vigorous exercise.[7] At the same time, some light physical engagement in daily routines could provide health benefits. Mayo Clinic researcher Dr. James Levine has described the value of light daily activities through his work on "non-exercise activity thermogenesis" (N.E.A.T.).[8] Author Dan Buettner has similarly observed how moderate daily activities lead to human longevity in centenarian *Blue Zone* communities.[9] In this book, the previously mentioned examples, including manual appliances, push-behind lawn mowers for smaller lawns, and even the collective use of manual tools and wind-up tools, could offer some reasonable N.E.A.T. activity. Generally, these examples offer broader ways to *reintegrate* some physical activity, and movement, into daily routines.

Certainly, not all manual solutions are going to appeal to everyone, but design and behavior change techniques could help advocate their benefits to a contemporary audience just as smartphones measure one's daily steps. How else can design encourage physical activity and movement while simultaneously reducing energy consumption? Additional design proposals seeking to answer this question are ever expanding. Some of these projects are provocations, and especially ambitious. For example, student designers Kai Mulligan and Noah Worden envisioned and built a concept: a fully functional gravity-fed ceiling fan. Electric ceiling fans are currently praised as an effective alternative or supplement to air conditioning in many climate situations. In this example, Mulligan and Worden explored gravity "batteries" for their project. Gravity "batteries" have been an obscure technology, first invented and used by Christiaan Huygens in 1656 as a means of powering a pendulum clock.[10] While gravity batteries were developed for different purposes into the 20th century, they were rarely, if ever, adapted for other kinds of products. In 2012, English inventor Martin Riddiford developed an LED light powered by a gravity battery, driving a dynamo.[11] Intrigued, Worden and Mulligan explored if this kind of past technology could be adapted to an appliance as large as a ceiling fan. Using a 30lb counterweight that must be hoisted about 5ft using mechanical advantage from ropes and pulleys, their Gravity Fan prototype demonstrated that light physical activity can provide up to 20–30 minutes of fan blade motion (along with a cool breeze) without being plugged in—perhaps long enough to take a short break and cool off or for use in off-grid purposes. Lighter materials than hard wood could provide further efficiencies.

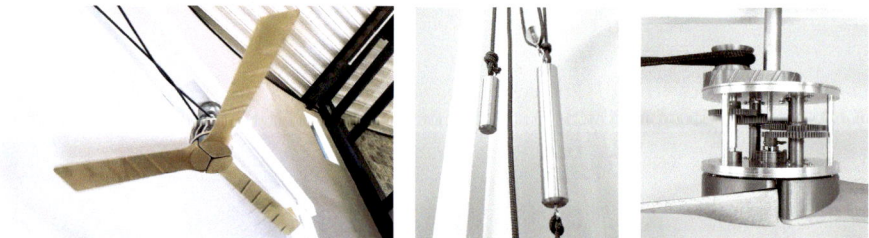

Figure 11.4
Above left: Gravity Fan, a concept for a human-powered ceiling fan using counterweights based on a 17th-century technology. Source: Kai Mulligan and Noah Worden.

Figure 11.5
Above center: The 30lb stainless steel weight that descends over 30 minutes to power the working prototype. Source: Kai Mulligan and Noah Worden.

Figure 11.6
Above right. A closeup of the Gravity Fan's mechanics. Source: Kai Mulligan and Noah Worden.

Principle 3. Design Longevity: Durability, Repairability and Adaptability

Volumes have been written in recent years about designing products to last longer to avoid the landfill and the costs of producing and buying replacements. In Chapter 6 (Return to Durable Design), this book shared numerous examples of industrial products whose lifespans were needlessly short. Disposable goods, designed to be used only once (or a few times), have become a troubling development since the late 20th century. Indeed, carryout packaging that is neither recyclable nor reusable has only added to this predicament, even in coffee shops where reusable cups used to be standard. Fear about virus transmission during the Covid-19 pandemic has only encouraged the use of single-use disposable products. These practices, of course, feed human convenience and generate more revenue for manufacturers, but they also create more waste. In contrast, durable products that once were maintained, reused or repaired have become increasingly short-lived and disposable, as was demonstrated in Chapter 6 with examples like the manual razor. On the other hand, efforts to reintroduce reusable, refillable containers that have long been disposable offer new opportunities to reduce single-use waste. Using more valuable, durable materials supports these efforts further, some which even develop desirable patinas over time, making these products more appealing, not less.[12] Designing products to last also helps engender enduring emotional bonds with their owners, thereby extending their lifespan further. Repairable products also offer a promising shift back towards design longevity, born of user frustration and enabled by right to repair legislation. Too often, electronic products from small appliances to the ubiquitous smartphone, quickly end up in landfills when they cannot be repaired or upgraded or because they are simply too difficult to take apart for recycling.

Finally, *adaptability* is another opportunity area where design can influence longevity. As was discussed in Chapter 6 with the child high chair step stool concept, the fleeting stage of childhood does not need to render a product useless once a child grows to sit normally at the dining room table. By making life-stage-based products adapt to new uses or pivot back to previous ones, they too can last longer. Generally, children's products that are designed to appeal to both kids *and* parents simultaneously can graduate to new practical uses as well.

Principle 4. Integrate *Alternatives* for Design in the 21st Century

If there is one broader objective this book hopes to achieve, it is to offer a fresh, meaningful design vision for the future to a wide audience: designers, engineers, inventors, educators and a broader public readership. Having presented design strategies that are distinct from those tied directly to the latest technology innovation, it should now be apparent that *alternative technology* can support society's current priorities, especially as more nations mobilize behind the UN's Sustainable Development Goals. In some cases, it has been shown how design can also help reinvigorate the *perception* of alternative technologies—to be appealing and relevant to users today. Hopefully, the audience of this book is also left with a more balanced view of technological innovation and its related energy and resource use—free from nostalgia or futurist hype. Further, there seems to be sufficient rationale for caution in the midst of what could be described as *Innovationist* rhetoric influencing the design fields, or an underlying predisposition towards the new—that what is new is inherently superior to what preceded. Often, industry promoters unwittingly reinforce an *Innovationist* mindset. ExxonMobil, for example, a leading producer of petroleum, offered the following mission statement in their 2019 *Outlook for Energy in 2040*:

> A house. A car. Lights at night and heat in the winter. A refrigerator to keep food fresh and a stove for cooking. A better education and a good job. Modern health care. Wireless communications. Technology and innovation. The freedom to focus one's daily activities on something more than mere subsistence. These are among the many benefits of modern energy . . . because energy is vital to our everyday lives.[13]

As this quote suggests, the promise and the *narrative* of innovation can be deeply seductive: connecting innovation, energy use and human fulfillment. Conversely, it is often the narrative of alternative technology, older and less glossy by comparison, that struggles for recognition next to such innovationist bravado. Just as Innovationism can be a foe of alternative and traditional technology, *compelling design* can be its best advocate. Some innovations, of course, are so gratifying that they would be hard to unseat, despite their costs. As mentioned many times in this book, it would be difficult to completely replace popular, energy-intensive innovations like AC, especially as temperatures regularly surpass 100 Fahrenheit (38 Celsius). The good news is that they do not have to be: the energy consumption and impact of this technology (and others) can be significantly reduced when *complemented by, and integrated with*, alternative technologies. What this will look like specifically in the built environment remains unknown and mostly undefined—it is up to the design fields to bring shape and life to these hybrid technological embodiments. As discussed in previous chapters, many well-known designers already have begun, and likely many more will continue to do so in the decades ahead. Looking to the future, designing a more resilient, sustainable built environment is certainly within reach—but it will require a multitude of new behaviors, tools, skills and technologies. Alternative technology should undoubtedly be among them.

Notes

1 Elferink, M. and Schierhorn, F. (2016) 'Global demand for food is rising. Can we meet it?' *Harvard Business Review*, April 7 [online]. Available at: https://hbr.org/2016/04/global-demand-for-food-is-rising-can-we-meet-it

2 Lampert, A. (2019) 'Over-exploitation of natural resources is followed by inevitable declines in economic growth and discount rate,' *Nature Communication*, 10 [online]. Available at: https://doi.org/10.1038/s41467-019-09246-2 (Accessed: August 2, 2022).

3 US Energy Information Administration (n.d.) *International electricity* [online]. Available at: www.eia.gov/international/data/world (Accessed: August 2, 2022).

4 Center for Sustainable Systems, University of Michigan (2021) 'U.S. Environmental Footprint Factsheet.' Pub. No. CSS08-08.

5 World Health Organization (2021) *Obesity and overweight*, June 9 [online]. Available at: www.who.int/news-room/fact-sheets/detail/obesity-and-overweight (Accessed: August 2, 2022).

6 World Health Organization (2020) *Physical activity*, November 26 [online]. Available at: www.who.int/news-room/fact-sheets/detail/physical-activity (Accessed: August 2, 2022).

7 Fanara, A., Clark, R., Duff, R. and Polad, M. (2007) 'How small devices are having a big impact on U.S. utility bills,' *Energy Star*. Available at: www.energystar.gov/ia/partners/prod_development/downloads/EEDAL-145.pdf (Accessed: January 25, 2022).

8 Levine, J.A. (2004) 'Nonexercise activity thermogenesis (NEAT): Environment and biology,' *Am J Physiol Endocrinol Metab*, 286(5), pp75–85. doi: 10.1152/ajpendo.00562.2003. Erratum in: *Am J Physiol Endocrinol Metab.* 2005 Jan; 288(1).

9 Buettner, D. (2008) *The blue zone: Lessons for living longer from the people who've lived the longest.* United States: National Geographic.

10 Milham, W.I. (1945) *Time and timekeepers.* New York: MacMillan. ISBN 0-7808-0008-7, pp. 330, 334.

11 Riddiford, M. (2012) *Gravity-powered electrical energy generator* [online]. Available at: https://patents.google.com/patent/US20120212948A1/en?assignee=DECIWATT+LTD&oq=DECIWATT+LTD (Accessed: August 2, 2022).

12 Materials like denim, leather, wood and metal can actually have more user appeal over time as they develop a patina.

13 ExxonMobil (2019) *The outlook for energy: A view to 2040* [online]. Available at: https://corporate.exxonmobil.com/-/media/Global/Files/outlook-for-energy/2019-Outlook-for-Energy_v4.pdf (Accessed August 3, 2022).

Index

Page numbers in **bold** refer to figures.